Eat My Dust

THE JOHNS HOPKINS UNIVERSITY
STUDIES IN HISTORICAL AND
POLITICAL SCIENCE

126TH SERIES (2008)

1. Georgine Clarsen, *Eat My Dust: Early Women Motorists*

Eat My Dust
Early Women Motorists

GEORGINE CLARSEN

The Johns Hopkins University Press
Baltimore

© 2008 The Johns Hopkins University Press
All rights reserved. Published 2008
Printed in the United States of America on acid-free paper
2 4 6 8 9 7 5 3 1

The Johns Hopkins University Press
2715 North Charles Street
Baltimore, Maryland 21218-4363
www.press.jhu.edu

Library of Congress Cataloging-in-Publication Data

Clarsen, Georgine, 1949–
Eat my dust : early women motorists / Georgine Clarsen.
p. cm.
Includes bibliographical references and index.
ISBN-13: 978-0-8018-8465-8 (hardcover : alk. paper)
ISBN-10: 0-8018-8465-9 (hardcover : alk. paper)
1. Automobile ownership—Australia—History—20th century.
2. Automobile ownership—Great Britain—History—20th century.
3. Automobile ownership—United States—History—20th century.
4. Women automobile drivers—History—20th century. 5. Women consumers—History—20th century. 6. Feminism—History—20th century. 7. Women—Social life and customs—20th century.
8. Nineteen twenties. I. Title.
HE5709.A6C586 2008
388.3'42082—dc22   2007053001

A catalog record for this book is available from the British Library.

Special discounts are available for bulk purchases of this book. For more information, please contact Special Sales at 410-516-6936 or specialsales@press.jhu.edu.

The Johns Hopkins University Press uses environmentally friendly book materials, including recycled text paper that is composed of at least 30 percent post-consumer waste, whenever possible. All of our book papers are acid-free, and our jackets and covers are printed on paper with recycled content.

To my parents,

Elizabeth Jeanne Clarsen-Lesueur
(1919–2003)
&
Bernard Jacobus Clarsen
(1919–2006)

Contents

Preface ix

Introduction 1

1 Movement in a Minor Key:
Dilemmas of the Woman Motorist 12

2 A War Product:
The British Motoring Girl and Her Garage 30

3 A Car Made by English Ladies for Others of Their Sex:
The Feminist Factory and the Lady's Car 46

4 Transcontinental Travel:
The Politics of Automobile Consumption in the United States 64

5 Campaigns on Wheels:
American Automobiles and a Suffrage of Consumption 86

6 "The Woman Who Does":
A Melbourne Women's Motor Garage 104

7 Driving Australian Modernity:
Conquering Australia by Car 120

8 Machines as the Measure of Women:
Cape-to-Cairo by Automobile 140

Conclusion 158

Notes 169
Essay on Sources 179
Index 189

Preface

Give a Girl a Spanner is one of the forgotten slogans of 1970s feminism, and I was one of the many girls in Australia who picked up a spanner (or wrench) and became a motor mechanic in those years. We female apprentices firmly believed that we were the first to take up such work, a misapprehension encouraged by stories in the press that celebrated women in male trades as pioneers moving into new territory. But every now and then there were disquieting moments. I remember being puzzled by the occasional conversation with older female customers who sought me out on the workshop floor to let me know they had done similar work when they were my age. Their pleasure in the memory of the work was palpable in their warm words of encouragement—but I had no way of making sense of the information they offered. Partly, it was my youthful arrogance that led me to assume that my generation represented an entirely new expression of femininity, but it was also true that I had been locked out of knowledge that should have been mine. I grew up in the postwar years, quite unaware of the rich history of women's active engagements with the mechanical arts. Not only was I ignorant of the facts of women's mechanical aspirations and achievements in previous generations, but there was no language available to help us understand a specifically female experience of technical competence or to articulate the challenges that we faced in an auto repair world that was firmly gendered male. We were only the newest generation of enthusiastic female mechanics, unknowingly *re*inventing relationships to technology—as though no technologically confident women had gone before.

It was not until two decades later, when I began interviewing women who had been motor transport drivers in the military services during World War II, that I came to understand how mistaken we had been. I found an earlier generation of women—my mother's generation—who, like us, had eagerly seized the pleasures and opportunities that the mechanical arts offered them. And through them I discovered something more astonishing: there had been an even earlier cohort of mechanical women, my grandmother's generation, who not only had worked as transport drivers during World War I but had opened motor garages, taxi companies, and rental car services in the interwar years. Female mechanics? Garage workers? Women's garages in the 1920s? How could it be that I did not know this? It was startling news, which impelled me to research the forgotten history of women automobilists in the decades surrounding World War I.

It has taken some ten years from that small start to the publication of this book, and there are many institutions and individuals who helped bring it to completion. At the University of Melbourne I was supported by an Australian Postgraduate Award and at the University of Adelaide by a postdoctoral fellowship at the Adelaide Research Center for Humanities and Social Sciences. A Fulbright Foundation Fellowship took me to the University of California, Berkeley, where I was an affiliate scholar at the Beatrice M. Bain Research Center for Gender Studies and a visiting fellow at the Transportation Institute. The Smithsonian Institution provided me with a postdoctoral fellowship to work in the archives of the National Museum of American History. An Australian Bicentennial Fellowship at the Menzies Centre for Australian Studies, University of London, and an Australian Academy of Humanities Traveling Fellowship supported my British research. The Ernestine Richter Avery Fellowship at the Huntington Library and Botanical Gardens, San Marino, California, and a research grant-in-aid at the Hagley Museum and Library, Wilmington, Delaware, allowed me to continue my American research. Finally, the writing was supported by an Australian Research Council postdoctoral fellowship at the Research School of Social Sciences, the Australian National University; a residency at the Center for Cultural Studies, University of California, Santa Cruz; and a research grant by the University of Wollongong.

My debt to many individuals is just as great. Patricia Grimshaw, Joy Damousi, and Anni Dugdale at the University of Melbourne helped me to formulate the very earliest versions of this project and have continued to offer help and encouragement. As the project became more ambitious and became a transnational project, a great many other people also provided invaluable support. Susan Magarey at the University of Adelaide; Caren Kaplan and Martin Wachs at the University of California, Berkeley; Roger White and Steven Lubar at the National Museum of American History; Roger Horowitz at the Hagley Museum and Library; Mark Patrick at the National Automotive History Collection, Detroit Public Library; archivists at the Benson Ford Research Center, Henry Ford Museum, Greenfield Village, Michigan; librarians at the Special Collections Library, University of Michigan, Ann Arbor; Patricia Jalland and Judy Wajcman at the Australian National University; Carroll Pursell, now at Macquarie University; Melissa Boyde and Amanda Lawson at the University of Wollongong; and the "Modernistas" at the

Australian National University—Ann Curthoys, Desley Deacon, Rosanne Kennedy, Alison Kibler, Jill Matthews, Ann McGrath, Fiona Paisley, and Angela Woollacott. I also offer my thanks to my colleagues in the Faculty of Arts at the University of Wollongong for their material support and encouragement. My debt to the many people who assisted me with each of the separate research studies is recorded in the notes to each chapter.

Ponch Hawkes offered me unfailing support from the beginning of the project and kindly helped me with selecting and reproducing photographs. Thanks to Belinda Henwood for her copyediting of the manuscript and writerly advice.

I have enjoyed the intellectual and personal companionship of a great many people throughout this research, including many who read early drafts of the manuscript. Special thanks to Barbara McBane, Julie Bishop, Ian Bracegirdle, Silvia D'Aviero, Lois Ellis, Kathy Gibson, Sarah Grimes, Sara Hardy, Donna Jackson, Margaret Jacobs, Viktoria Lopatkiewicz, Julie McInnes, Richard Mohr, Barbara Nicholson, Inez Rekers-Lesueur, Kathy Robinson, Amanda Smith, and Ginette Verstraete.

An earlier version of chapter 3 was published as Georgine Clarsen, "'A Fine University for Women Engineers': A Scottish Munitions Factory in World War I," *Women's History Review* 12, no. 3 (2003): 333–56. Parts of chapter 6 were published as "'The Woman Who Does': A Melbourne Motor Garage Proprietor," in *Sapphic Modernities: Sexuality, Women, and National Culture*, ed. Laura Doan and Jane Garrity (New York: Palgrave Macmillan, 2006), 55–71. Parts of chapter 7 were published as "Tracing the Outline of Nation: Circling Australia by Car," *Continuum: Journal of Media and Cultural Studies* 13, no. 3 (1999): 359–69. Material from chapter 8 will be published as "Machines as the Measure of Women: Colonial Irony in a Cape to Cairo Automobile Journey (1930)," *Journal of Transport History* 29, no. 1 (2008). I thank each of the publishers for permission to republish.

Finally, I would like to thank Bob Brugger, senior acquisitions editor at the Johns Hopkins University Press, his assistant Josh Tong, and copy editor Elizabeth Gratch for their help and support in seeing this project into publication.

Eat My Dust

Introduction

On a summer night in 1888 in Mannheim, Germany, Bertha Benz became the first person to drive an automobile on a social journey, the kind of trip we now recognize as a perfectly ordinary thing to do in a car. With her teenage sons, Eugen and Richard, Bertha pushed the prototype car built by her sleeping husband out of his workshop, far enough from the house so he would not wake when they fired it up, and set out to visit her mother on the longest trip ever taken in an automobile. For two years Karl Benz had carefully test-driven his invention in the streets of Mannheim, but Bertha was impatient, determined to prove that the machine was ready to go into production. In spite of several breakdowns, the three drove over sixty miles across the steep hills and rough roads of the Black Forest to arrive at "Mutter's" house in Pforzheim late that evening. The journey would have taken more than two days by horse and buggy. Bertha's bold move proved to be a public relations triumph, and the press, which had been skeptical about the strange machine, began to take Karl's work seriously. Bertha's actions signaled the transformation of what had been an item of industrial production, the internal combustion engine, into a safe and manageable object of domestic consumption.

A little more than a decade after Bertha's inaugural family automobile journey, cars were well on the way to becoming an accepted part of upper-class domestic life in industrial societies. This book is about the ways that women continued to play a key role in the processes that propelled cars into our social lives, despite the fact that their desires to be recognized as competent motorists were frequently rebuffed. It presents a series of stories about women in the United States, Britain, Australia, and British colonial Africa, from the early 1900s to the 1930s, who aspired to be motor mechanics, taxi drivers, long-distance adventurers, racing drivers, automobile engineers, and just knowledgeable everyday motorists. Car ownership in those early years was generally confined to a privileged few. Regardless of their social and economic advantage, however, these women were obliged to do a great deal of extra work to secure a place for themselves in the practices and conversations that surrounded automobiles. The very term *woman motorist* indicated that they were supplementary to the main game. These studies of particular women in specific historical moments and precise locations show how they had to exert themselves in order to be accepted as competent motorists, far beyond men of their race and class, their fathers, brothers, husbands, and male friends.

Men quickly claimed automobiles as a prized masculine technology and presumed to define the terms under which they should be adopted into social life, and thus my challenge has been to present women's love of cars as more than just a pale imitation of theirs. While the motoring industry welcomed women as consumers, the idea that they might develop an authoritative relationship to cars—becoming capable drivers, knowledgeable purchasers, happy tinkerers, professional mechanics, or creative designers—was a different matter. Manufacturers and their agents frequently used the slogan So Simple That Even a Woman Can Drive It throughout the first decades of the century and well beyond, in an attempt to reassure hesitant men, as much as apprehensive women. Time and again, male judgment, often in the form of "anti–woman driver" humor, confidently announced that the most that could be expected from women was a timid and uninformed response to machinery, rather than a mastery of it. Wherever they turned, aspiring women motorists found transparently partisan definitions of automobile technology that worked to place them on the margins, even as they were invited to become consumers of it. The result was that motoring women composed their words and actions with an eye to the largely unwelcoming environment they encountered.

Each chapter presents a story pieced together from a variety of sources—newspaper accounts, motoring magazines, privately published travelogues, books now out of print, letters, journals, advertising images, and interviews as well as from archival collections. Every account considers a distinctive moment or expression of female automobility, turning an apparently commonsense story of women in general's purported lack of aptitude and interest in machinery into an exploration of how actual women produced, affirmed, and performed their love of cars in those early years. The studies show how the unwelcoming climate they encountered provided the basis for a collective identity from which women were able to contest their exclusion. They illustrate how women's actions were firmly stamped by the possibilities for change they found at hand, as they cobbled together whatever resources they could employ to their advantage. Each study therefore is steeped in the concerns and conditions of its own national context and historical moment.

Rather than acquiescing in a milieu that constructed them as strangers on foreign territory, early women motorists produced their own enabling stories and material sites, creating the conditions for their participation in worlds otherwise denied them. They wrote novels, autobiographical books, advice manuals, and newspaper columns that asserted women's capacity for a knowledgeable affinity with machinery; they opened taxi services, driving schools, automobile clubs, and motor garages, where they created a sympathetic environment within which they could enjoy themselves, learn skills, and earn a living; they dreamed of becoming engineers; they reinvigorated political life by incorporating automobiles into their suffrage campaigns; and they traveled long distances—across a continent or around the world—to declare their belongingness in the citizenship of the road. In their actions and words early women motorists claimed a degree of worldliness as technological actors, challenging notions of female technological ineptitude and automotive naïveté.

Together, these small-scale stories of particular women in specific places tell a much larger story of how automobiles were incorporated into twentieth-century women's aspirations for major social change—for independence, mobility, meaningful work, and pleasurable lives. Although their actions reflected a particular national context, the women in these studies were acutely aware of other female motorists around the world. Thus, when women refused to accept the unfavorable terms of sexual difference they encountered in the motoring world and rejected its exclusions in their own lives, they framed their actions in terms of social changes that were occurring in many other places. The transnational focus of this book reminds us that, just as people and cars routinely crossed national borders, so too did the ideas, inspirations, and public debates that expressed and shaped women's desires for change.

At their most basic levels each of the stories show how women's engagements with automobiles have been much messier than supposed by the commonsense division between men as naturally competent and avid technological actors and women as naive and hesitant ones. They record women devising ways, specific to each decade and the circumstances they found themselves in, to counter the judgments that so ungenerously classified and surveilled them as bad drivers, inept mechanics, and timid travelers. In addition, the studies show these women recognized that the sharp line drawn between men and women was a fiction that contributed to the masculinity of the technology by actively producing the very outcomes it posited.

For all their social privilege motoring women who sought to handle automobiles with knowledge and confidence were frequently the objects of male censure. Motoring women were acutely aware of being forced to respond—in ways that did not exacerbate the difficulties they experienced or spoil their pleasure—to the power that men had to define them as trespassers onto masculine territory. The women's writing reveals that they keenly observed the extra demands that a masculine presumption of entitlement placed on the most ordinary of women's everyday actions. They articulated how men's license to view women's practical exertions with an authorizing gaze could construe their actions as strange or inappropriate. In 1908 mother and daughter Minerva and Vera Teape drove their two-cylinder Waltham buckboard from Denver to Chicago, attracting a crowd of "helpful" men and boys whenever they stopped—especially when they crank-started the car or made repairs. And twenty years later, when Kathryn Hulme and her friend Tuny drove their Dodge roadster across America, they too reported that they were forced to resort to repairing their car in uninhabited places because they had "many experiences with the 'assisting' man camper who thinks that when a woman gets anything more complicated than an egg-beater in her hand, she is to be watched carefully."[1]

The writing of early women motorists reveals their frustration at how beliefs about masculinity and femininity, instantly imposed with an ascribing glance, worked to prescribe what they might be permitted and encouraged to do or be inhibited or precluded from attempting. Their words drew attention to the practical effects that the masculine terms of the technology had on their lives, as they carefully analyzed how sexual difference was a powerful register for ascribing capacities and bequeathing effects. From the smallest action of holding a tool to more ambitious desires to establish a motor garage, manufacture an automobile, or drive across a continent or across the world, the women voiced how they experienced the coercive weightiness of sexual difference. First, they clarified how women's alleged estrangement from automobile technology meant that it required extra effort for actual women to hold a tool or embark on independent travel—it imposed a degree of effort that could in fact be much more burdensome than the demands of the activity itself. Second, their words point to how women's estrangement from technologies, produced under such disadvantageous terms, garnered an easy relationship with the technological for men, thereby strengthening the very categories of sexual difference they were seeking to place into question.

These studies highlight the productivity of automobile technology when it comes to sexual difference. They portray automobiles not simply as value-free items that were delivered fully formed by gender-neutral designers and manufacturers to a generic end user but as malleable objects whose production continued long after they left the factory floor. The very nuts and bolts of automobiles and the cultures that surrounded them, by inviting some people into a privileged relationship and seeking to limit others, expressed competing conceptions of social life and invoked unequal ways of moving and being. By placing women's investments and women's interests at the center, rather than accepting men's experiences as the norm, this book locates automobiles as sites of struggle over claims to authority and entitlement in the field of a technology, in which both the meanings of automobiles and notions of gender were at stake.

When they placed their hands on the steering wheel and in the engine compartment, sped around the racetrack, adopted more functional clothing, traveled far from home, used automobiles to earn their living, and wrote about what those new experiences meant to them, motoring women were declaring that they were active agents at the heart of twentieth-century life. In their words and actions they consciously located themselves as part of a progressive project of inventing and articulating new versions of what it meant to be a modern woman. The studies suggest that their interventions into a technological field should be considered a form of political action, as they were performed with the avowed intention of making notions of women's technological incompetence less valid, less viable, and less inevitable. Furthermore, as women redrafted notions of femininity through their love of cars, they were engaged not only in struggles over what it meant to be female in the twentieth century but also in debates about what an automobile was, how it might be used, and in which directions the technology might move in the future.

Automobile technology emerged in industrialized societies just at the moment when a variety of formerly silenced social groups were vigorously proclaiming their right to full citizen status, entitled to an equal place in public life. Enthusiasm for the social changes of the twentieth century—the powers, pleasures, competencies, and freedoms that new constituencies anticipated—often became bound up with, and expressed through, the consumption of mass-produced goods and new services. During the late nineteenth century privileged women relished their newly won freedom to walk alone through commercial streetscapes, to shop in department stores, to ride on public transport, and to pedal bicycles. Their daughters welcomed with equal enthusiasm the opportunity to drive automobiles and fly airplanes. But it was not only privileged women who avidly embraced commodity consumption as a means of expanding their lives in the early decades of the twentieth century. Working-class women, particularly in the United States, were taking no less delight in the new patterns of consumption, mass-produced fashions, nickelodeons, cinemas, and urban amusement parks.

From the very first years of the twentieth century African Americans, too, were investing liberatory hopes in new forms of commodity consumption. Defying the racial segregation of the so-called open road, they welcomed automobiles according to their own needs and desires. Rather than viewing cars as objects of privatized consumption, as manufacturers promoted them, some black activists attempted to redefine cars as a collective resource for sidestepping Jim Crow. As early as 1905, the African-American press was reporting that blacks in Nashville had formed a consortium to buy automobiles, which they planned to run as shuttle buses in opposition to the city's segregated trolley cars. In Alabama, Nate Shaw, who had been an activist in the Sharecropper's Union, recalled that by the early 1920s some African-American sharecroppers in his neighborhood were buying cars for both work and pleasure. Shaw learned to drive his brother's car in 1923, and by 1926 he was the first African American he knew to have bought a new Ford—so that his sons could go out courting and have fun in it, he declared.[2] With the growth of a black urban middle class in the 1920s, African Americans were increasingly becoming car owners, able to mount legal challenges against the racial discrimination they encountered on the road. They responded with optimism to the liberatory potential of automobiles, establishing hotels, resorts, insurance companies, and travel agents to cater to black motorists and travelers in those early decades. For African Americans, as well as other groups whose access to public space was limited, automobile consumption was much more than an economic activity; it was incorporated into their aspirations for social and political change—for active, modern citizenship.

More than just a contribution to automobile history, the studies in this book work to locate women automobile consumers within twentieth-century changes to class, gender, sexuality, race, nation, and empire, thereby adding thickness to our understanding of the linkages between them. Privileged women motorists' personal ambitions for expanded lives were associated with broader movements and events that were central to the development of twentieth-century modernity—be it in suffrage campaigns; the expression of new sexual identities; the expansion of the middle class; the emergence of mass production, mass consumption, and the mass media; or unfolding national identities and a new colonial order. By considering female motorists as avid and creative modernists, self-declared "modern women" able to direct the terms of their lives, the stories offer a new perspective on the personal, emotional, and bodily sources of twentieth-century sensibilities. The book, which begins with questions about what it meant for a woman living in the United States to want to hold a repair tool in her hand in the first years of the century, ends with a consideration of female motorists in the world of changing colonial regimes in the interwar period.

Women's location at the margins of automobile technology, caught in the dilemma of being simultaneously welcomed as consumers but disparaged as incompetent technological actors, provides a new register through which to understand how gendered differences were experienced, contested, and reworked in everyday interactions. The stories highlight how women—as a distinct category of identity, both socially defined and personally experienced—acquired new meanings under particular historical circumstances. They demonstrate how commonsense, dualistically conceived categories of "men" and "women" are not simply static descriptors for natural biological differences but are social classifications, always contested and undergoing constant change. And in any struggle to effect change, language—the terms, categories, metaphors, and narrative forms available to construct meaning—constitutes a critical element.

When the aspiring female motor mechanics of feminism's "second wave" in the 1970s tried to understand differences in men's and women's relationships to technologies, the terms of analysis we produced were *negative sex-role stereotypes* or *oppressive gender roles*. We optimistically declared that it was merely irrational stereotypes about men and women's social roles, rather than biological differences between the sexes, that prevented women from reaching their true potential as technological actors. The new language of sex-role differences developed by feminists was a powerful tool for expressing the socially determined (and thereby alterable) meanings that had become attached to natural differences between men and women. Although at the time it was enabling, that broad brushstroke language also served to limit our analysis of the deeply contradictory issues of sexual difference. In the first instance the terms perpetuated a rather naive orientation in which "oppressive social roles" were simply imposed upon women from an unspecified "outside." This approach failed to consider how women themselves might be active agents and protagonists of their destiny—both inside and outside of the notions of sexual difference they sought to challenge. Second, while sex-role analysis emphasized differences in social roles that were amenable to redefinition, at the same time it quarantined from analysis how bodily differences between men and women were understood and experienced. The language of sex roles assumed that how men and women lived as embodied beings was fixed and beyond social definition, rather than subject to social shaping. Ultimately, the simplicity of the terms proved to be less than useful because they did not go very far to illuminate how women might go about negotiating the difficult circumstances they found themselves in. The terms were inadequate to the task of analyzing the subtle nuances of sexual difference that women who sought to move into masculine territory daily faced.

Thirty years on, feminist analysis is moving toward a more sophisticated understanding of the power differentials between men and women, one that seeks to take greater account of the diversity of women's powers and social locations. In that spirit this study depicts women motorists as both powerful and powerless, engaged in a modern project of imagining and producing themselves in new terms. Rather than portraying early female motorists as simply discovering, declaring, or revealing their "true selves" in the face of negative sex-role stereotypes or oppressive social structures that had unjustly thwarted them, early motoring women's words and actions are here presented as creative acts. Women motorists' creativity lay not so much in trying to express or bring to light a more authentic femininity in the face of exclusionary practices and stereotypes that repressed them but in imagining and producing new female identities in ways that raised questions about cultural understandings of gender. The studies reveal how early women motorists were engaged in a project not simply to sidestep repressive stereotypes or combat unjust practices that placed them at a disadvantage but, much more productively, to shift commonsense notions of a fixed division between men and women, always and already given in nature.

While women motorists' actions were only occasionally characterized as feminist in their own time, in this book I have presented them as part of a political quest, a struggle to alter women's lives in fundamental and far-reaching ways. The concerns of motoring women were usually distant from what has been recognized as overtly feminist campaigns to secure suffrage, legal rights, access to education, political representation, or employment opportunities. Nevertheless, their determination to redraft the everyday terms of femininity—such as women's purported physical weakness and lack of coordination, their timidity and nervousness, their lack of mechanical aptitude or sense of direction—constitutes a different kind of politics and one that is rightfully situated within the history of feminist activism. Women motorists' contestation of the limitations they faced is an important but neglected element in the profound shifts in ideas of masculinity and femininity taking place throughout the twentieth century. These women made it clear that it was not the purely mechanical arts, narrowly defined as the acquisition of new fields of knowledge and skills, that presented their major difficulty. Acquiring mechanical skills, they frequently intimated, was the least of the challenges they faced. More important, they were grappling with fraught questions of how to fashion themselves into females who could feel properly at home within a domain in which being male had been construed as the norm. Their attempts to create the category of the technologically adroit woman spilled over into a much bigger project, that of redrafting the very ideas of masculinity and femininity.

Motoring women's challenges to the restrictions they experienced were expressed in a different sphere, however, from that of more conventional political campaigns. These women were not primarily claiming particular rights and responsibilities or new forms of political and legal recognition, though that was sometimes part of their aim. More than many other kinds of female activism, they were above all emphasizing their desire to do physically exciting and challenging new things. Their struggles centered on the engagement of their whole bodies in a world of practical activity—the "ordinary purposive orientation of the body as a whole toward things and its environment."[3] What their bodies were able to do, encouraged to do, and permitted to do were motoring women's key concerns. Their struggle lay in how to apply their female bodies to tasks that had been designated as proper to, and even constitutive of, the active male body. Through their determination to reinvent and publicly inhabit their differently lived bodies within a masculine automobile world, these women identified the female body as a site of struggle.

As the editors of a special 1997 edition of *Technology and Culture* identified, the technological reshaping and redefining of the human body is an emerging field of study for which conceptual tools and language are still being developed.[4] In approaching these stories, my intention has been to contribute to that project by recognizing the embodied nature of automotive knowledge and emphasizing the ways that early women motorists clearly identified it as such. Not only did early motoring demand a whole-body engagement with unfamiliar machinery and new physical skills, but for women motorists it also threw into question received beliefs about appropriate female bodily comportments. These women's writings carefully articulated the complexity of their positioning as active and knowledgeable motorists and hinted at the depth of emotionality attached to the powers and competencies ascribed to men and women's bodies. The studies give due weight to their detailed focus on the embodied nature of their motoring experiences. They foreground not only the bodily pleasure the women derived from their newly acquired accomplishments but also the intractability of ideas of sexed bodily competencies, so acutely experienced by women, whose access to technological worlds had been fraught.

The words and actions of early motoring women record their attempts to construct viable new identities that reworked their experience of sexual difference in the most basic terms: the minute, lived, dailiness of the "arts of using the human body," as one of the first theorists of everyday bodily practices characterized it.[5] As the women coherently articulated, precisely the same action, whether it be operating a lathe, using a wrench, driving a car, traveling far from home, changing a tire, having an accident, or burning rubber just for fun, had quite different meanings depending on whether it was performed by a male or a female body. Furthermore, those same yet profoundly different actions, performed in public, were taken to be integral to the properly constituted maleness or femaleness of those bodies. The questions women motorists asked about sexed machines, sexed tools, sexed roads, sexed hands, and sexed bodies drew attention to how they experienced the everyday, lived fullness of differences between men and women within the technological field of motoring.

Women motorists posed questions about what covert work was being done when diverse bodies with widely varying capacities were repeatedly classified into just two kinds—male and female. They struggled to devise a language by which to explore and express some of the effects of that ascribed polarization, so deeply experienced but only partly a matter of words. Most important, they explored the process of the emergence of alternate bodies, differently "engineered" by new machines, new practices, new discourses, new habits, and new pleasures. In doing so, they raised questions about the history of bodies, asking how bodily powers and representations of them might change over time. As they formulated and enacted their own versions of mechanical pleasure, these women remind us that what women did, their noncompliant bodily acts, as much as what they argued or reasoned, should be considered another form of feminist politics.

That does not mean, however, that these stories of motoring women are about pure feminist heroines, if such people exist. Rather, they reveal women's challenges to men's power in the field of motoring to be contradictory, self-interested, and sometimes even counterproductive. These were women fully imbricated in all the power relations of their era, and, at the same time that they contested their gendered exclusion, they exploited their race and class privilege with tactics that were often just as inconsistent and ungenerous as some of the restrictive practices and unwelcoming environments they sought to transform.[6] In their ambition to be included as the beneficiaries of technological modernity, they were open to any terms that were available to them and hypersensitive to whatever gaps and points of leverage they might use to their advantage. Their sense of identity was constructed in opposition to the claims of others, such as working-class men and people of color, who they frequently believed to be less modern and less deserving than themselves.

Rather than the standard story of social progress in which women have been steadily accepted into car culture, the studies presented here provide a much more uneasy history. Early women motorists' success in creating a publicly acknowledged and honorable place for themselves during the first three decades of the century was not sustained into the 1930s. As automobile ownership spread beyond the privileged few, women drivers faced increasing hostility and ridicule from male drivers. Even though the number of women drivers grew rapidly in the succeeding decades, most of them did not have the resources or the access to public forums that had enabled the socially privileged motorists of the first three decades to press their claims. The result was that the gains that some women made in one generation were wound back and all but forgotten by the next. In remembering those early gains, I offer this book as a contribution to the larger project of telling rich and nuanced histories about the antecedents of our contemporary automobile culture. In the twenty-first century it is more important than ever to complicate a linear narrative of technological progress by exposing the unruly dimensions of the past. Neat and coherent stories of progress overlook the ways in which the emergence of automobile culture, which has come to dominate industrial societies, was discontinuous and counterintuitive. A less carefully sutured narrative troubles the fiction that the future is already fixed, and a proliferation of unexpected stories about the past can give us courage to believe that there may be new ways to move.

CHAPTER 1

Movement in a Minor Key

Dilemmas of the Woman Motorist

Women drove automobiles from the time that the first models came into production, and male commentators noted with surprise that they were avid "operators." In 1896 the first American motoring journal, *Horseless Age*, recorded the enthusiasm of stylish women motorists on the streets of Paris, and by 1900 it was reporting on female drivers in London and New York. The journal revealed that there were as many as fifty women who regularly drove in Chicago—many of them without having paid the license fee. In Britain motoring magazines such as *Autocar* and *Car Illustrated* were also recording the presence of women motorists at the turn of the century. Not only were women enthusiastic consumers of early automobiles, but they also began to publish their motoring experiences almost as soon as they took the wheel. Women produced novels, how-to manuals, magazine articles, and regular columns in the motoring press, in which they strongly advocated women's capacity as "chauffeuses," as women drivers were then awkwardly called. Although their numbers were far fewer than male motorists and their voices muted in comparison to the noisy and careless declarations of masculine entitlement, women's involvement in the emerging technology is unmistakable—once we expect them to be there and once we learn to read and listen differently. It is important to acknowledge and recall their active presence, not simply to correct the record. Women were acutely aware of cars as a technology of masculine privilege, and from their marginal location they tell us something new about the experiences of early motoring, details both ecstatic and mundane. Their writing fills out our understanding of the origins of our present car culture.

It has been quite forgotten that in the first few decades of the twentieth century the fledgling mode of travel demanded a degree of accommodation on the part of both passengers and operators that amounted to a major shift in the relationships between people and machines. Most automobile histories make much mileage out of how horses were scared by the appearance of those new vehicles. Many also record comic stories of the ways peasants and farmers—hicks and rubes—first mistook them for noisy and fume-belching apparitions from hell. Most, however, entirely neglect to note how automobiles were also quite unfamiliar to the privileged men and women who were first able to buy them. In fact, both men and women had to acquire new habits and skills carefully and self-consciously in order to become an "automobilist."

Automobile owners had to do a great deal of work to have the fully modern experience of driving, rather than be chauffeured, and, as cars entered upper-crust domestic life, they had to undertake a sort of intensive training in the technological arts. The proliferation of books for the "amateur motorist" in those early years and the popularity of the first driving schools attest to the strangeness of the experience. When the Boston YMCA opened an automobile school in 1903, it was overwhelmed by demand from both wealthy car owners and men hoping to find work as chauffeurs. Similarly in Britain, numerous driving schools were opened in the early years of the century targeted to the lady and gentleman owner-driver as well as to men who wanted to become professional chauffeurs.

Learning to feel at home with automobile technology required immense shifts of perspective in a class of people who had previously been largely insulated from such practical matters. Electric cars were the easiest to adapt to and demanded the least amount of activity to operate. Even so, a lengthy article by Hiram Percy Maxim in 1898 revealed how carefully that new skill needed to be demonstrated to the first-time female driver. Given their longer antecedents, steam-powered automobiles were already the subject of official regulation, and Anne Rainsford French, the daughter of a prominent doctor in Washington, D.C., had to qualify for a steam engineers' license before she was permitted to drive her steam Locomobile in 1900. When it came to internal combustion engines, there were generally few licensing laws, beyond the payment of a fee, but the challenge was to learn how to be comfortable while "sitting on top of an explosion," as early gasoline automobiles were sometimes characterized. Those profound class shifts in habits, areas of knowledge, and comportment styles instigated by the new machines were far from trivial. They set the scene for tremendous changes in perceptual and bodily experiences that would reverberate throughout the twentieth century. *Town and Country* magazine reflected in 1902 on the emergence of the "expert amateur chauffeur" and noted the incongruous conjunction of extreme wealth and mechanical interest. It had become the latest fad for wealthy men to act as their own machinist, and it was not unusual to come across a famous son of a multimillionaire, "lying prone upon his back, lost to public view, underneath a mass of costly French machinery, investigating for himself the cause of some stoppage in transit."[1]

Unlike the passive mode of railway travel, driving required an active, whole-body engagement that propelled early motorists, both privileged men and women, into direct contact with what had been industrial machinery. It forced them as nonspecialists to acquire some of the actions, habits, knowledge, comportment, and styles of clothing that had formerly been the domain of those professional engineers, working-class men, and farmers who worked with stationary engines. Even so, the gasoline engines fitted to automobiles were somewhat different. They were higher revving and needed to be coaxed into performing through a range of revolutions per minute and under variable load conditions, rather than at the slow, steady beat of the stationary engine. Motorists had to adjust the ignition timing and fuel-air mix to the driving conditions. They needed to develop an ear for the note of the engine and the rhythm of the drive train and be able to judge the precise moment for a gear change. They had to diagnose mechanical faults, handle unfamiliar chemicals, learn to use tools for routine maintenance and roadside repairs, beware of tire "sideslip," and always remember to keep the oil pressure pumped up. It was essential that they be prepared to become greasy, dusty, wet, windblown, muddy, and thoroughly shaken about. Furthermore, they had to consider it a sport or pleasurable activity (or at least admirably modern) rather than déclassé or just sheer hard work.

In short, a whole new technical, bodily, and social vocabulary was being composed as that privileged, new social category emerged—the private automobilist, male or female. Much more than simply adopting a new form of transportation, becoming a motorist demanded the conscious acquisition of an entirely new repertoire of bodily disciplines. "Cultivating an acute sympathy for the mechanism" was how it was frequently expressed. That close affinity was thought to be a necessary condition for enjoying the pleasures of controlling a piece of complex machinery that delivered range, speed, and a gratifying sense of power. Electric cars failed to measure up on all those counts, and even men promoting them admitted that "nobody has the slightest desire to drive them himself."[2] Given electric vehicles' limitation in range and speed, men considered them far too simple and dull, but, defining themselves as the natural inheritors of mechanical aptitude, they offered electric cars as the perfect vehicle for women's circumscribed mobility and lack of technical talent. "No license should be granted to a woman, unless, possibly, for a car driven by electric power," boomed one commentator in 1909. "The natural training of a woman is not in the direction to allow her to properly manipulate an automobile in case of emergencies."³

Even though manufacturers and their agents carefully promoted electric cars to the female market, particularly in the United States, most women preferred gasoline cars. Like men, women wanted the pleasures and conveniences of mobility, control, power, range, and affordability—along with the comforts of electric technology transferred into gasoline cars, such as electric lighting, self-starters, and weatherproof, enclosed bodies. That preference women expressed for gasoline automobiles should be seen as a significant act, one with far-reaching consequences. Their resistance to the parlor-like comfort of electric cars—silent, simple to drive, and suitably "lady-like," but much more expensive and far less exciting than cars with internal combustion engines—signaled their intention to refuse Victorian notions of women's dependency, refinement, and restriction to a world separate from men. In adopting gasoline cars, women were choosing to embrace the physically challenging and socially heterogeneous life of the open road. It meant that they were obliged to perform a great deal of extra work, far beyond that of men of their own class and race, so that they could acquire the kind of affinity with machinery that early motoring required.

Hilda Ward, a twenty-year-old artist living on Long Island, New York, was among the first women to produce a written account of her fraught journey into motoring knowledge. Her book, *The Girl and the Motor*, provides an invaluable description of a young woman's painstaking and self-conscious struggle to acquire mechanical confidence.⁴ Published in 1908, *The Girl and the Motor* recorded not only Ward's desire to drive automobiles but her determination to understand, maintain, and repair them as well. She wrote in a lighthearted and breezy style, which mixed self-deprecating humor with ruling-class confidence. Her book included sketches of her repairing her car that bore captions such as "Dependent on No Mere Man," or "My Busy Day." In spite of the flippant air she adopted for much of the book, Ward's account of sexual difference in automobile technology was a serious one, as she poignantly articulated the ways that her journey toward technological competence was very different from the one her friend and rival, "Mere Man," could assume.

Like many young women enthusiasts, Ward's finances only allowed for modest motoring ambitions. She had spent a summer mastering the quirky, single-cylinder, two-stroke motor on her small dinghy and began to think about buying a secondhand car. She studied motoring journals and technical literature, pored over catalogues, haunted automobile showrooms, and bombarded agents with endless questions. She even imagined that she could revolutionize the industry by designing a one-cylinder engine that could "receive an impulse every half turn as in a steam engine" (40). Most of all, Hilda Ward counted her hard-earned savings over and over and worked to silence her family's objections to her motoring plans.

Once Ward found the right car, she was determined to maintain and keep it on the road herself. It was not just a matter of keeping the cost down, she declared; she also did it for the sheer pleasure and sense of power that such self-reliance brought her. Ward sought out professional "mechanicians" for a direct line to mechanical knowledge and used the novelty value of her enthusiasm to good effect. When the main bearings needed relining ("re-babbitting"), she arranged to place herself in the hands of her local "Master Mechanic" and, wearing her oldest clothes, drove her car to his workshop. With the engine dismantled on the floor, she was able to "study the transmission and see the wheels go round to my hearts content." To the amusement of other customers, Ward spent the day happily scraping cylinders and grinding valves under the direction of the Master Mechanic. With the car reassembled at the end of the day, she drove home "resolved that henceforth all work would be done by me, and the mechanics do only that which home tools and a good tennis arm were powerless to effect" (46–47).

Ward relished the many hours she spent in her home garage and the Master Mechanic's workshop. She expressed surprise that her friends could "never learn that I enjoy monkeying with screwdrivers and wire-cutters and oil cans at least as much as running the car" (71). For her automotive knowledge brought changes in her relationship to the technological world, which she expressed in bodily terms. Frequent breakdowns, she wrote, have induced in her "abnormal qualities of caution and alertness of eye, ear and nose" (39). A "horrid grinding, crushing, snapping sound within the motor" elicited an instinctive, physical, response: "a mad snatch for the brakes and the spark thrown off as rapidly as my fingers could work" (50). Hilda Ward's growing bodily adroitness and mechanical ease was further bolstered as she came to appreciate the rather shaky foundations underpinning the technological mastery that some men claimed. She spent much of the book railing against her male friends' assumption of mechanical superiority, carefully documenting whenever they dispensed confident, but erroneous, mechanical advice. It was galling to her that men of her own class could presume to belittle her "life as an engineer" and that professional chauffeurs, her social inferiors, could lord it over her even when their knowledge was rather sketchy.

The edge of bitterness that invaded her otherwise upbeat narrative ensured that Ward's playfulness was always slightly forced, her enjoyment deliberate and measured. She had to work hard at her pleasure, her status as an outsider already guaranteed by the earnestness she was obliged to adopt. In her determination to produce herself as a mechanically skilled subject, she inevitably lacked Mere Man's relaxed assumption of natural ownership and unquestioned entitlement, whatever his actual competence. Reluctantly, she had to acknowledge that, no matter what her enthusiasm, dedication, or aptitude, there always remained a lack of fit between her aspirations to be a real engineer like the Master Mechanic and the social constraints imposed on her female body. "I am only a girl after all, and what right do I have to be fooling around with engines anyhow?" she was forced to concede (98). Most pointed was her apparently simple question, arguably the central theme of *The Girl and the Motor*. Noting the "quiet but very effectual way" that the Master Mechanic handled repair tools, she asked, "I wonder why no tool ever looks really at home in a woman's hand?" (99). Her question went straight to the heart of women's experience of sexual difference in the domain of technology and arguably remains there even today.

Hilda Ward posed the question, of course, about particular tools. It was not babies' bottles or brooms (or even sewing machines, typewriters, or electric lighting) she was thinking of but, rather, specific "men's tools," the wrenches and screwdrivers needed to repair the kinds of machinery that men loved. For all her determination Ward had to concede that she had no answer to offer. She accepted that she could never become a real automobile engineer and extracted some comfort by elevating her amateur status to a fine art. Ward told proud stories of impressing strangers with her unexpected competence and took great pleasure in helping motorists she found stranded on the side of the road. Her book ends with one such story, to which I will return.

Ward's references to her unnamed friend, Mere Man, inverted men's joking references to women automobilists in the popular automobile journalism that proliferated during the early years of the century. Their stories worked to consolidate the social terms within which the technology acquired its meanings. A typically light-hearted article, entitled "Why Women Are, or Are Not, Good Chauffeuses," published in the New York sporting magazine *Outing* in 1904, when cars were still a recreational accessory for fashionable society, perfectly illustrates the climate within which women were obliged to approach automobiles. In it three men puzzle over a new social conundrum—how to make a romantic approach to the coolly confident young female motorists in their country club, intimidating determinedly competent women rather like Hilda Ward. In the course of their discussion and in spite of the evidence before them, those fictional men happily concluded that very few women were likely to become competent motorists. "Nerves and impulsiveness," they agree, "are precisely the two things that make women fall down as chauffeuses." Women's very natures, they conveniently concurred—their impulsiveness, foolhardiness, indecision, and inability to understand machinery—prevented them from mastering a car; such command, the men agreed, "amounts pretty near to mastering herself."[5]

Because the idea of a woman mastering herself seemed a remote possibility and perhaps even a contradiction in terms, the men were able to recover their sense of natural entitlement. They declared "a woman's hands must be trained to the uses of the appliances of her car, as the hands of a pianist are trained." Women drivers needed to "act without the special direction of the brain and almost without the aid of the eye—indeed almost mechanically" (159). The young men's confident assertion of sexual difference transposed into a new technology served as advance notice that women's claims to competence in such a prized field were to be closely scrutinized. For, in spite of the fact of women's active presence as motorists and notwithstanding their commercial value as consumers, women were compelled to perform public tests not applicable to men.

Privileged men could assume an identity of technically competent consumers, should they want to, with a minimum of fuss. Everywhere they turned, they could find commentators such as the motoring editor of the *London Sphere*, a passionate advocate for the "amateur motorist," who declared in 1907 that "any man of intellect could become the master of his car within a week."[6] There was a place reserved for them whenever they might choose to take up the invitation. In fact men had actively to renounce the assumption of mastery in order to evade its claims—and many of them did. But women who wished to assert their technological accomplishment were subjected to unwelcoming scrutiny. As Hilda Ward discovered, even before they put a tool in their hand, women were already positioned as strangers on foreign territory and were forced to respond to the ungenerous assessments of their abilities and entitlement that were so frequently expressed in public and private forums.

In 1904, even earlier than Hilda Ward's book and the same year that "Why Women Are, or Are Not, Good Chauffeuses" was published, American journalist Mrs. A. Sherman Hitchcock began to write motoring columns for and about women, though, like Ward, she is now rarely remembered. In her first article for *Motor: The National Magazine of Motoring,* Hitchcock wrote enthusiastically about the "fascination" and "vast pleasure" of driving, rather than merely sitting in the passenger seat. In her column, "A Woman's Viewpoint of Motoring," she described the feelings of power and exhilaration that women experienced when they discovered how easy it was to drive—"when the ponderous car begins to move and the motor seems a living, breathing thing, respondent to your slightest touch, easy to control and simple to manipulate."[7] Mrs. Hitchcock continued writing regular magazine columns "for the motoring woman and about her" for at least ten years, never deviating from that optimistic assessment of the transforming possibilities of a woman-machine affinity. Confidence was the key to Mrs. Hitchcock's formulation. Learning the road rules, developing a cool head under pressure, and acquiring a deep knowledge of "the mechanism" were her mantras.

Hitchcock characterized mechanical expertise as a bodily, sensory attunement to machines, which came with time and practice. Using a musical image quite different from automated fingers on the piano suggested by the men in the *Outing* story, Mrs. Hitchcock declared that the motorist needed to "educate her ear to the normal sound of the engine and thus become able to discover, amid the whizzing of the wheels, humming of gears, and throbbing of the motor, any wrong note, indicative of trouble, before it seriously affects the working of the machine" (19). Hitchcock's imagery invoked a pleasurable, knowing, whole-body engagement with the mechanism and the potentialities it contained, and she boldly pronounced that women's capacity to so harmonize their bodies to the sounds and rhythms of automobiles was much the same as men's.

Other privileged women besides Hilda Ward and Mrs. Hitchcock, inside and outside the United States, articulated similarly optimistic assessments of women's capacities for a knowledgeable affinity with automobiles in those years. As early as 1902 in Britain, Lady Jeune was writing about the "enchanting sensation" of flying along country roads in an automobile. In the same year the English "New Woman" author, Mrs. Edward Kennard, published *The Motor Maniac*, a novel built around one wealthy woman's obsession with owning, driving, and maintaining a variety of automobiles. The novel's energetic heroine, Mrs. Janet Jenks, unlike her staid husband, Algernon, reveled in the new technology and loved nothing more than spending whole days on the road and in the repair shop with Snooks, her "motor man." A well-known motorist herself and the author of regular women's motoring columns in the fashionable London magazine *Car Illustrated* at the turn of the century, Mrs. Kennard devoted much of her novel to an analysis of the new bodily sensations that driving delivered as well as to long descriptions of the mechanical intricacies of motoring and careful assessments of the merits of the various French, German, and British motorcars that a buyer could choose. Her book *The Motor Maniac* portrayed automobiles as immensely appealing to forward-thinking women who relished the experiences that modern life could bring, while many of their men remained stuck in the slow routines, rhythms, and habits of the past.

There were many other women who similarly advocated the suitability and pleasures of automobiling for women in those early years. In 1904 Gladys Beattie Crozier wrote a feature for the *Ladies' Realm* on "Motoring for Ladies," illustrated with numerous photographs of fashionable women who were perfectly at home driving, adjusting, and repairing their cars. Other exclusive magazines such as the *Tatler, Car Illustrated, Daily Graphic,* and *Gentlewoman* all ran regular motoring columns for women from the first years of the century. In 1906 journalist and popular English writer Eliza Davis Aria published a wryly humorous book, *Woman and the Motor Car: Being the Autobiography of an Automobilist,* about her conversion to the "fine, careless rapture" of speed. The book ranged across mechanical tuition delivered in catechism style ("What is a petrol engine?"), her adventures on the road, and the foibles of the rich automobile set while on holiday in Brighton's Hotel Metropole. Baroness Campbell von Laurentz, a Scotswoman who had also been contributing to British motoring magazines since the turn of the century, published *My Motor Milestones: How to Tour in a Car* in 1913. Her book was devoted to the practical aspects of long-distance travel, such as how to repair punctures, the special pillow she placed behind her so that she could reach the pedals, and the boxes she had designed to store her luggage. In Australia, too, there was an active female voice in early car talk. The principal automobile magazine, the *Australian Motorist,* included a column for women called "A Woman's Point of View," by "Minerva," from its first edition in 1908 until 1913, when it became "Women A-Wheel" by "Artemis," published until 1919.

It was Dorothy Levitt's *The Woman and Her Car: A Chatty Little Handbook for All Women Who Motor or Who Want to Motor,* published in London in 1909, that enjoyed the greatest circulation of all the women's motoring books. An Englishwoman named as "the premier woman motorist and botorist of the world," Levitt had raced in numerous motorboat and automobile events in Britain and Europe with "skill, courage and cool judgment" since 1903. "I never take a mechanician with me," she declared, "I attend to my cars myself . . . and believe in being absolutely self-reliant."[8] Her *Woman and Her Car* was a how-to manual that portrayed gentlewomen in command of their own machines. It was illustrated with many photographs of Levitt, dressed in a linen artist's smock and leaning over her small one-cylinder De Dion car, completely at ease with a tool in her hand. The photographs, designed to demonstrate how repairs could be made without jeopardizing either her femininity or class privilege, were given encouraging captions such as "The Adjustment of the Foot-Brake Is a Matter of Seconds" and "It Is a Simple Matter to Remove a Faulty Sparking Plug." *The Woman and Her Car* was reviewed at length in Britain, the United States, and Australia shortly after it was published, and Levitt was quoted in women's motoring columns on how women all over the world used automobiles for pleasure and profit. Even though there were significant national differences in automobile production and consumption in those three countries, women's writing conjured up imagined cosmopolitan communities of female motorists and initiated transnational conversations about the ways that women could enjoy and profit from the new technology.

Unlike Ward's *Girl and the Motor*, Levitt's book was not an autobiographical account of her motoring experiences, nor did she ask questions about how tools might come to look at home in women's hands. *The Woman and Her Car* was a practical manual, addressed to the "active gentlewoman . . . desirous of being her own driver and owning her own car" but who was nervous or who imagined that it might be too difficult to understand "the mechanical details of automobilism."[9] In specifically addressing women who did not dare to motor, Levitt's book is redolent with the extra work that women needed to perform to produce themselves as competent motorists. She articulated a highly conscious awareness of women as a special category, always and already standing in a tentative relationship to automobile technology. "Take your time and get in sympathy with your motor as you would the horses you drive or ride. Gain confidence slowly. Once you have confidence in yourself the battle is nearly won" (47). "Train your ear to distinguish the slightest sound foreign to the consistent running of the engine," she urged in terms that were reminiscent of Mrs. Hitchcock's advice to her American readers (52). And, like Hilda Ward, Dorothy Levitt drew heavily on notions of inherent class superiority to bolster privileged women's claims to technological mastery. She declared that any woman who wanted to learn how to drive and manage a motorcar could do so as quickly as a man. Hundreds of women had done so, and there was many a woman "whose keen eyes can detect, and whose deft fingers can remedy, a loose nut or a faulty electrical connection in half the time that the professional chauffeur would spend on the work" (86–87).

Much of women's automobile writing was characterized by a refreshing and brash opportunism, the writers attempting to create room for themselves within that technology of masculine privilege. There were always plenty of commentators such as Montgomery Rollins, who in 1909 put on record his conviction that "the natural training of women is not in the direction to allow her to properly manipulate an automobile in case of emergencies" and "at the first sudden appearance of danger a woman is not to be compared to a man in level-headedness."[10] But women countered those pronouncements, making creative use of any resources available to them to put forward their claim that they, too, could be admirable motorists. Some even followed novelist Mrs. Edward Kennard's lead and made bold claims that women were naturally better motorists than men and temperamentally more suited to the particular strains that modern life brought. Joan Cuneo, who was famous throughout the East Coast of the United States as a racing and rally driver until the American Automobile Association banned her from competing against men, was one prominent woman who expressed great confidence in women's special adaptability to motoring. As she put it in 1910, "It is a curious fact that if she goes at motoring seriously, woman's natural intuition puts her in closer touch with her car than a man seems to be able to get with his." A woman could acquire the "feel" of the mechanism and detect something out of adjustment more easily than a man, she thought, and had the capacity for driving more gently and with a greater "delicate technic"—a talent that gave women particular pleasure and satisfaction as motorists.[11]

Most audacious of all were some women's claims that it was precisely the "nerves and impulsiveness" from which the heroes of the *Outing* article derived comfort that made women good motorists. If, as late-nineteenth-century vitalist theories declared, the quickening pace of life and bewildering proliferation of new sensations were the source of the nerve fatigue, known as neurasthenia, one of modernity's bodily effects, then some women proposed to turn their putative weaknesses into assets that gave them distinct advantages in adapting to modern life. Far from being shocked by the new, they welcomed what it might bring them, strategically employing medical discourses to declare that it was precisely their feminine weakness and nervy vitality that made them particularly well adapted to the demands of motoring. Flightiness and a lack of capacity for deep thought, they argued, ensured women's special fitness for that most trying of all modern experiences, city driving. Melbourne's "Artemis" argued the case in her defense of women drivers in 1915 in her monthly column, "Women A-Wheel." She asserted that a woman's "superficial" senses were more keenly developed than a man's, making her more aware of the many things happening around her, while a man was inclined to concentrate laboriously on just one thing. Men might be able to act coolly and with more considered logic, but a woman's quick, alert consciousness of her surroundings stood her in good stead in an emergency. She could, "nine times out of ten, act more promptly—by intuition and in obedience to her finer feelings."[12]

That line of argument was not new. Mrs. A. Sherman Hitchcock had advanced it in a number of her motoring columns and enlisted male experts who extended its applicability to the even more glamorous technology of aviation. In July 1911, for example, Mrs. Hitchcock quoted at length the flamboyant airman, Capt. Thomas S. Baldwin, who declared, "The bird-woman is going to excel the bird-man in the field of aviation." It was women's many weaknesses, he argued, that made them

"A woman's natural intuition puts her in close touch with her car." Joan Cuneo, one of New York's premier racing drivers, 1910. Courtesy of the Detroit Public Library, National Automotive History Collection.

specially adapted to aviation. He listed among them a lack of curiosity, which enabled women to obey directives without question; a facility for attention to detail; mental flexibility brought about by their tendency to jump from one idea to another; and the capacity to keep a cool head because of their susceptibility to be distracted by trivialities such as their hairdo in moments of danger.[13]

On offer were fresh terms for women to represent their aspirations, and those wanting to imagine different futures seized—opportunistically and on the run—such backhanded affirmations of their special adaptability to technological modernity. Women turned apparently disparaging characterizations of themselves as naturally childish, absorbed by the momentary and unaware of the mechanical gravity of their play, into resources through which they might conceptualize their experiences as drivers and capture a legitimate place within that technology. Flexibility, vitality, nerviness, superficiality, and spontaneity—all characteristics that men widely attributed to women at the turn of the century—could be turned on their head to provide a discursive tactic that allowed some women to resist the unwelcoming environment they encountered and to locate themselves as properly at home in the driver's seat.

The terms on which even privileged women were able to take part in early automobiling were circumscribed. They were acutely aware of the limitations placed upon them and had little choice but to negotiate their way carefully around them. Motoring women were obliged to struggle, in ways that men were not, with fundamental issues of differences between what bodies were able to do, were imagined to do, were allowed to do, and were encouraged to do, in a domain in which female bodies were largely dismissed as being out of place. Even as they tried to constitute themselves as active, technologically savvy subjects, women could not help but feel the pressure that their status as objects of curiosity or unsympathetic observation and restraint placed them under.

The stakes were high, for there was little room for fudging when it came to early automobiles, and symbolic claims of entitlement or competency did not count for much when things went wrong—as they frequently did. The novelty of early motoring made female drivers highly visible in open cars on the road, where they were exposed not only to dust and weather but also to increased social surveillance. In the terms of the time, one did not drive in a car but on it, and sitting high up behind the steering wheel inevitably propelled motorists into something of a public spectacle. That, of course, was the desired effect and a large part of the pleasure, but their enjoyment of it could easily and without warning flip in the opposite. A mechanical breakdown far from home became a test not only of a motorist's resourcefulness and knowledge but also the kindness of strangers, and early women motorists inevitably operated under a heightened sense that the pleasure, power, and prestige conferred by their automobilic independence were always rather fragile. The public scrutiny women experienced meant that they could instantly be defined as "out of place" when things went wrong. They might even risk physical danger, as suggested by the frequent speculation about whether women driving without the company of men needed to carry a pistol.

Women's writing reveals an acute consciousness that they were being carefully observed when they were out driving and that consequently they, too, needed to watch themselves closely. It is not surprising, then, that clothing—one of the most visible markers of class and sexual difference—was a primary concern to early female motorists, just as it had been for female cyclists a decade before. "The all-important question of dress," Dorothy Levitt called it in her book, devoting a chapter and several photographs to what driving women should wear. Lady Jeune had warned that it was fraught territory, declaring that it "affects women very deeply."[14] Eliza Aria, describing her recent conversion to automobilic obsessions and recalling her previous scorn for motoring fashions, agonized in her book, *Woman and the Motor Car: Being the Autobiography of an Automobilist,* "How shall I dress the part?" She dedicated three chapters to her dilemma. And, in keeping with women's increasing role as consumers of mass-produced goods, almost all women's motoring columns of the prewar years featured a large section on fashion, which carefully described and illustrated new fabrics and styles for motoring coats, hats, rugs, and even jewelry. Great attention was paid to the novel cut of a jacket, a weatherproof fabric such as "pongee" or "asquascutum," or the design of a motoring bonnet that could be cunningly transformed into something more suitable for everyday wear.

Women's insistence on the importance of dress, far from being merely frivolous or indicating their special susceptibility to consumerist ideologies, was an integral part of their promotion of that "delicate technic" that Joan Cuneo had declared them to be particularly capable of. The pleasure and care female motorists took over their consumption of clothes was central to the new sensibility they were working to create between their bodies and machines. Fine distinctions were no minor matter to those in the know, and experienced motorists were precise in their approach to the question of clothing, giving careful advice to novices. For repair and maintenance work Dorothy Levitt recommended a butcher-blue or brown overall of washable linen, similar to the simple smocks worn by female artists, and gloves that could be cleaned in petrol. As for what to wear when driving, she suggested practical clothing that came as close as possible to everyday fashions. Levitt emphatically declared that under no circumstances should women motorists wear lace or anything "fluffy" (25).

At issue was how to acquire an admirably modern, capable, and adventurous look without inviting social ridicule. It was a fine balance for the motoring woman to achieve, for she had to negotiate a path between respectability, practicality, and being fashionable in a variety of social situations, without veering toward either tasteless ostentation or overly utilitarian clothing that would be considered unfeminine on a woman of privilege. Being acutely conscious of these subtle issues, the primary value that female motorists emphasized was practicality, rather than a strict adherence to fashion. But both practicality and fashion, like "the modern" itself, have no fixed referents and may be interpreted in many ways. When Christobel Ellis raced a stripped-down Arrol-Johnston in the first women's event at the Brooklands racing circuit in Surrey in 1908, she had little choice but to wear a long skirt, though the car's open body presented major difficulties for keeping it in place. Male racing drivers during the Edwardian period were able to adapt working-class men's fashions into raffish new styles to suit the demands of motor racing, but

Ellis did not have the same freedom in presenting her active female body. She adopted the dangerous solution of hobbling her legs while she was speeding around the racetrack by tying a cord around her calves to keep her skirt in place.[15]

Certainly, motoring women, in emphasizing the importance of suitable clothing to their new identities, were not trying to do away with fashion, as they sometimes liked to claim. Struggles over fashion were critical to women's strategies for developing their new identities. Motoring women's clothing constituted a visual declaration of their claims to new female ways of comporting themselves, new competencies, and new modes of independent movement. Their preference for particular fashions was designed to express the stand they wished to make. They were working to produce a physical display of entitlement, in which the way they appeared constituted a declaration of social change, enacted in a new technology and indicated through bodily signs. If their act was the public performance of their competence and independent mobility, then their clothes were the way they advertised that act.

As Christobel Ellis's dilemma of how to race in a long skirt suggests, precisely the same bodily action, whether it was racing a car, holding a tool, changing a tire, re-babbitting (re-lining) the bearings, or simply doing everyday driving, had quite different meanings according to whether it was performed by a male or a female body. And, furthermore, those precisely the same yet profoundly different actions seemed inseparable from the properly constituted maleness and femaleness of those bodies. So, when a woman crawled under a car to make a repair or adjustment, or when a woman was driving alone, crank-started a car, entered an automobile race, or was involved in an accident, a whole constellation of meanings was called into being that did not apply to precisely the same action performed by a man. Similarly, getting dirty had quite different meanings depending upon whether the dirty body was male or female; young or old; black or white; ruling class, middle class, or working class—and, of course, what kind of dirt it was.

At stake were questions about how men's and women's bodies were differently marked, properly authorized to assume particular ways of comporting themselves, to wear certain kinds of clothing, to perform specific actions, to circulate at will, to observe with a certain stance of humor and detachment, to occupy public space, and to claim easy access to privileged languages and sets of knowledge. Early women motorists felt those differences acutely—instantly conferred by an appraising glance. Their writing suggests how they experienced the distinct spheres of activity typically ascribed to men and women in the field of technology as a cramped and confining space. Hilda Ward knew she was as good a mechanic, or as bad a one, as the neighbor with whom she had a friendly rivalry. She could not help but see, however, the entirely different consequences that her female body bequeathed in her relationship to technological things, and she railed against its injustice.

Knowing that precisely the same activity had different meanings depending on what type of body was performing it ensured that Hilda Ward could not move with the same unselfconscious intentionality or playful irreverence that Mere Man enjoyed. She was obliged to create laboriously a space that gave her a chance to enact the competent subject she desired to be. It was a reality that pressed on her, a force toward containment, that could remain quite outside the consciousness of her male contemporaries, free as they were from such weighty injunctions. Another way to state her question about male tools and female hands is how, after she had put herself through the process of developing particular kinds of knowledge and skills, could she, too, inherit the power that inhered to them? What extra work need she perform to be able to stamp her female body with the marks of its authority? How else might her body become? How plastic could her flesh be?

Hilda Ward's solution to her dilemma was to invoke her class and race privilege. Like other early women motorists, her claims to competence were buttressed by crude sideswipes at working-class men, those other aspiring motorists she could categorize as patently less modern and less deserving than herself. At that time wealthy motorists were loudly complaining about the power of chauffeurs, who were a new class of expert servant that could not be easily placed within established social hierarchies. Chauffeurs were thought to hold ideas "above their station" and often refused to show the deference expected of them. They provoked a great deal of resentment among automobile owners, endlessly elaborated in motoring journals and books, and, within that widespread culture of complaint, privileged women such as Ward worked to construct points of leverage for their own advantage. They proposed that, as wives, mothers, and daughters eager to be behind the steering wheel and in the engine compartment, women were well placed to take over the functions of the household chauffeur. "Sometimes I think my reminiscences should be entitled 'How to be happy without a Chauffeur,'" Ward wrote, ending her book with a familiar moral story about the danger of being dependent on professional chauffeurs (57).

In that story Ward expressed crude disdain for a black chauffeur she had found stranded by the side of the road, unable to diagnose the fault for his white employers. From the very earliest days African-American chauffeurs were a controversial part of automobile culture; the New York press reported attempts by some white chauffeurs to expel them from the profession by sabotaging the vehicles they drove. "'Queering' the Negro Driver," one headline named it.[16] Hilda Ward had much to gain by "queering" both black and white professional chauffeurs. Just as Mere Man advanced his technical superiority by squeezing out women drivers, Ward sought to promote her own interest by bringing into play notions of class- and race-based superiority at the expense of those demonized workers of the day. "Costly machines should not be entrusted to the detrimental hands of unscrupulous hirelings!" she declared in terms that were echoed by men and women of her own class in the United States and beyond (58). In placing herself as the rescuing hero of a party stranded by their black chauffeur's ignorance of mechanics, Ward confidently assumed that her readers would agree that her white, ruling-class hands would naturally do a better repair job than those of a black chauffeur — even if those hands were also female.

Such class- and race-based strategies to monopolize technical expertise had only limited success for women such as Hilda Ward, as they briefly stepped into places that hired chauffeurs might otherwise have occupied. Any advantage they derived was quickly lost, as both automobiles and people's relationships to them became configured in new ways. In the United States especially, cars rapidly escaped the bounds of extreme class privilege, and mechanical competence became attributed to sex, race, and class differences in new ways. The private chauffeur disappeared from all but the wealthiest households, and the "amateur driver" became the "owner-driver" and eventually just a "driver." The rapid democratization of the technology in the decade after her book was published and the changing class, sex, and racial codifications of twentieth-century consumption quickly made her attempts to call upon inherent class and race superiority in mechanical matters anachronistic and even laughable. Ward's desire to redefine the power imbalances that adhered to notions of sexual difference within automobile technology — to evade the sharp line drawn between men and women in that field — were confounded by her determination to draw other lines of difference even more firmly.

Ward's ambition of finding professional employment through her love of automobiles, her ardent wish for a "life as an engineer," was, however, a thoroughly modern desire that resonated with other privileged women throughout the industrialized world. Her book associated automobile technology with wide-reaching transformations in women's lives and demonstrates how the distinctive social category of the modern young woman, most often associated with the changes brought about by World War I, was actually taking shape well before the war. Other women were similarly investing great hopes in the new technology, many achieving more success in earning a living through their automotive accomplishments than Hilda Ward managed. Upper- and middle-class women in Britain, the United States, and Australia created new social identities and opportunities for employment, which were sometimes expressed in the most flamboyant and theatrical of ways. Couched differently in each national context and each decade, they were identities built on women's newly found ease with urban street life, on their delight in the bodily freedoms they were embracing, on the increasing importance of commodity consumption, as well as on their painstakingly acquired familiarity with a modern technology.

CHAPTER 2

A War Product

The British Motoring Girl and Her Garage

In the first decades of the twentieth century young women's paid employment became a feature of upper- and middle-class family life; indeed, it was integral to the expansion of the middle class throughout the century. Women who were lucky enough to have access to automobiles were well positioned to take up emerging forms of employment as licensed taxi drivers, professional chauffeurs, garage proprietors, drivers in the military services, or auto mechanics in private business. It was not only financial independence and meaningful work that attracted women into professional work as motorists in those years but also the mobility and engagement in public life that automobiles offered. Although they were rarely well paid, numbers of privileged young women declared the work to be much more interesting and glamorous than office work. Dorothy Levitt, the author of *The Woman and Her Car: A Chatty Little Handbook for All Women Who Motor or Who Want to Motor* (1909), was among the first of the British upper-class women to become a professional motorist. She had been secretary to Selwyn Edge, the managing director of the British motor manufacturer Napier, until the company sent her to a mechanical training course in France in 1903. A talented and daring driver, she became one of their foremost works drivers in competitive racing and sales demonstration as well as a prominent commentator on women and motoring.

When Sheila O'Neil became one of London's professional taxi drivers in 1908, she attracted a great deal of interest on both sides of the Atlantic.[1] "I think the profession of motor-driving is a most suitable one in every way for women," she declared to the press, and for the next two decades other young women echoed her sentiments, using whatever means available to turn her optimistic statement into reality. In her mid-twenties and the daughter of a military officer, O'Neil was a trained nurse who had traveled extensively through India and Africa. She had served for two years in the Boer War and had been caught in the Siege of Ladysmith. Economic necessity may not have been her primary motivation in becoming a taxi driver, as within two years she was again the focus of press reports, this time announcing her plan to fly across the Irish Sea in a biplane of her own design. Instead, O'Neil cited her war service as well as her class entitlement and imperial privilege to support her actions. Her public stand for employment was as much a bid to expand the social possibilities open to her as a traveled, upper-middle-class and cosmopolitan woman as it was to earn her living.

Forced to operate privately from a large motor garage because Scotland Yard had refused to issue her with a license to work from a taxi rank, O'Neil wore the King's and the Queen's medals, awarded for her work in the South African war, on her sable fur motoring coat. She displayed her medals as affirmation of her civic status, a stamp of royal recognition that underpinned her claims to a continuing and visible place in public life. They proffered proof, furthermore, that she possessed the "steady nerves" thought necessary to drive motorcars through the congested streets of London. By the end of her first day Sheila O'Neil announced that she already had regular bookings to take ladies shopping and had been engaged by a doctor for two hours every morning to take him on his rounds. Her dramatic stand was recorded in admiring press articles that could not resist making jokes at her expense. Reporters declared that her training as a nurse would enable her to provide first aid to anybody whom she ran over. They named her as the first woman taxicabist, though a great many more women would be called that name throughout the following decades and no doubt had been called that before. Certainly, there had been women cab drivers in the horse-drawn era. In 1891 British papers noted the death of Betsy Collins, a widow from Gloucestershire, who had owned and driven the van omnibus between the village of Anst and the city of Bristol for more than thirty years, and in 1897 the London Cab and Omnibus Company announced plans to employ twenty-five women hansom cab drivers.[2]

More than a year before Sheila O'Neil had created a stir in the press, a British automobile journal published a letter sent to the Motor Drivers' Employment Agency, noting that any of their readers who assumed the domain of the chauffeur was "safe from the encroachments of woman" should be prepared for a "rude shock." The young woman seeking work as a driver wrote, "I can do all running repairs, and put tyres on, and am willing to make myself useful."[3] What struck the editor of *Autocar* as remarkable was how she wrote "in the most matter-of-fact kind of way, as though there were nothing at all extraordinary in her application, and she was not at all conscious of introducing the slightest innovation." Her letter caused the editor to accelerate toward an alarming future, "not merely of feminine chauffeurs, but of female mechanics and mechanical engineers as well, and mere man will be driven to seek fresh outlets for his powers." His alarm, facetious though it may have been, hinted at the ways in which automobile technology implicitly staged sexual difference, offering a view of automobiles as a technology that delineated a distinct, but now contested, boundary between the proper activities of men and women.

References to women's desires for motion—to "leave the office chair for the sunlit stretches of the open road, and the whirr of the motor," as reporters liked to put it—were found in articles and advertisements scattered throughout the British press in the prewar years. They appeared with increasing frequency as World War I began to change the opportunities for women's work and as enterprising women established businesses that catered to women's desires to become expert motorists. The exclusive magazine *Queen: The Ladies' Newspaper and Court Chronicle* noted in 1914 that there were several flourishing ladies' motor businesses in London and the provinces. Miss Alice Hilda Neville had established a driving school and repair garage at Worthing, near Brighton, in 1913. Two years later the Honorable Gabrielle Borthwick had opened her Ladies' Automobile Workshops in Brick Street and Grantham Place, Piccadilly, advertising her business with the slogan "Women Trained by Women." In the same year Miss C. Griff, billing herself as a consulting engineer who offered expert advice on automobile, electrical, and mechanical engineering matters, opened a workshop on Dover Street, Mayfair. Miss Griff provided mechanical repairs and courses in "elementary and advanced motor mechanism," with evening lectures on industrial and factory training available at "especially low fees" to prospective women industrial workers. In addition, there were at least three women's garages in the wealthy suburb of Kensington in the early war years: Miss Amelia Preston, one-time chauffeur to Mrs. Pankhurst, was offering courses in "motor driving and running repairs" from her workshop in St. Mary Abbott's Place; Miss Nora Bulkley, a former instructor at Miss Preston's motor school, was providing a similar service at the Warwick School of Motoring; and by 1916 the Women's Volunteer Reserve ran a garage in Cromwell Mews managed by Mrs. Charlesworth, who trained women to pass mechanical and driving tests on ambulances and commercial vehicles.

"Mechanism, driving and running repairs taught" was the typical language found in advertisements, indicating that early driving went far beyond simply learning to operate a motor vehicle but required consumers to understand the new machines and to be able to service and repair them as well. The courses were very

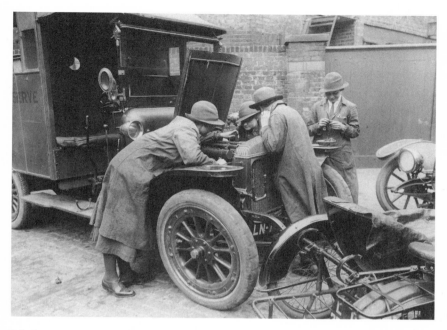

Mechanics training at the Women's Volunteer Reserve Garage, London, 1916.
By permission of Getty Images.

expensive, geared to the privileged woman of independent means. Twelve driving lessons, which included mechanical instruction, cost as much as five guineas. At a time when many female industrial workers, even under higher-paid wartime conditions, were earning little more than one pound per week, the price of an extended apprenticeship at Gabrielle Borthwick's Ladies' Automobile Workshops was a formidable twenty-one pounds, with a shorter course costing six pounds five shillings, still a substantial amount. Once qualified, a Women's Legion driver in France could expect to be paid only one pound fifteen shillings per week, of which almost three-quarters was deducted for board, not to mention the cost of uniforms and kit. A woman entering private service in England as a companion-chauffeuse or a van driver for a commercial business earned slightly less.

The growth in women's automobile training during the war years was underpinned by the acute shortage of skilled drivers and mechanics as Britain mechanized its military services and drafted qualified men. Private garage workshops were short-staffed; department stores such as Harrods and government departments, including the postal service, began recruiting female drivers; and women's voluntary services—such as the Women's Emergency Corps, the Women's Legion, the Women's Volunteer Reserve, and various ambulance units such as the Scottish Women's Hospital, the First Aid Nursing Yeomanry, and the Allies Field Ambulance Corps—all trained and mobilized female drivers. Gladys de Havilland, sister to the famous airman and airplane engineer, published *The Woman's Motor Manual: How to Obtain Employment in Government or Private Service as a Woman Driver* (1918), which provided a thorough guide to the new forms of employment open to women.[4] For the price of three shillings women could read about what kind of work was available—from taxi driving, private service, sales and demonstration, and military motorcycle riding to motor plowing—as well as find a list of motor schools, learn where to apply for employment, the hours and rates of pay they could expect, licensing laws, as well as the basics of driving, mechanics, and repair work.

The sudden visibility of uniformed, competent women on British roads was a spectacle that drew much commentary during the war years. It appeared in government propaganda, press reports, fiction, advertising, and entertainment. In 1919 the Clincher Motor Tyre Company, for example, published advertisements showing two uniformed women driving a department store delivery van on a busy city street, and a new feature of the Royal Naval, Military and Air Forces tournament at Olympia was a wheel-changing competition for female drivers in the military services. Numerous patriotic photographic exhibitions as well as articles and photographs in the popular and feminist press showed women service drivers doing repair work in the French mud, steering ambulances through devastated landscapes, or driving high-ranking military officers to the War Office in London. But, while the unexpected emergence of competent female motorists in the war years has been frequently noted, the broader context of women's claims to automotive knowledge has received little attention. The antecedents to women's wartime transport work, the ways that British women collectively organized to train and encourage each other, and the determination by many women to continue as professional drivers and auto mechanics after the war was over deserves more consideration. How did it happen that there were so many women already skilled and eager to take up motoring work at the beginning of the war? What were the terms through which motoring women represented their professional aspirations, and by what means did they attempt to maintain their precarious positions behind the wheel and their hands in the engine compartment in the postwar years?

Surprisingly, these questions have been neglected until very recently in both automobile histories and in histories of women. Perhaps that neglect represents a failure in historical imagination, as privileged women's ambitions to work as professional drivers or in auto garages are now rather mystifying. Viewed through the subsequent ubiquity of car ownership and the downgrading of the mechanical arts, taxi driving and automobile mechanics generally rank as low-status, male, working-class occupations. Current middle-class judgment renders those early women's aspirations strange, making their actions all but invisible in historical accounts. But, for the British women whose hopes and identities revolved around automobiles and who derived pleasure, status, income, and mobility from their newly found knowledge, the immediate context was not that of downward social mobility. Instead, they expected the automotive exclusivity and class privilege that had obtained before the war to continue in the postwar period.

What we now recognize as the coming democratization of automobility through mass production and mass consumption was not yet a feature of the British motoring scene, as it was in the United States. In 1924, for example, there were seventy-eight residents per motorcar in Britain, while the figure in the United States was already an astounding seven residents per car. It was not until 1963, more than forty years later than the United States, that Britain attained the ratio of seven persons per car.[5] So, at the forefront of privileged British women's ambitions in the early decades of the century were the gendered dimensions of motoring within elite circles. The technology's status as highly valued masculine knowledge, its promise to open out new ways for women to circulate freely through public spaces on terms similar to men of their own class, and its scope for opening out new kinds of experiences are crucial to understanding upper-class women's interest in professional auto work in Britain during the years surrounding World War I.

As they climbed on, in, and under their dispatch motorcycles, ambulances, lorries, taxis, delivery vans, and touring cars, British women were exploring and exploiting the transformative possibilities of a new technology, through which they aspired to create themselves as new kinds of women. Apparently played out in the economic fields of business and employment, it was at the same time a struggle to redefine female identity at the level of everyday bodily actions and comportments. Fashion and styles of dress, far from being peripheral to their aspirations, were central concerns. Military-style clothing had become all the rage for well-to-do patriotic women during the war, though not without a great deal of public controversy. Exclusive sporting and military outfitters, such as Burberry or Derry and Tom, produced "National Service Suits" for well-off women engaged in active war work. Feminist journals such as *Common Cause* and magazines such as *Graphic* carried advertisements for corduroy bib and brace overalls in "useful shades" at about nineteen shillings, and a khaki waterproof suit—coat, breeches, leggings, and matching hat—cost just over two pounds.

Simply getting into masculine clothing released tremendous feelings of pleasure. When Vita Sackville-West first put on breeches, gaiters, and boots for her land work in 1918, she wrote, "I went into wild spirits; I ran, I shouted, I jumped, I climbed, I vaulted over gates, I felt like a schoolboy let out for a holiday."[6] Racing driver Christobel Ellis, no longer obliged to hobble her legs to maintain her modesty while she raced her stripped-down Arrol-Johnston, became head of the motor transport branch of the Women's Legion during the war, organizing female motor drivers for the army. In that new role she took a great deal of care to design appropriate clothing styles for women's active war work.

Changes in women's fashions were significant, as privileged women's clothing became detached from Edwardian associations of leisure and conspicuous consumption and became identified with patriotic sentiments, robust physical activity, competence, modernity, and professionalism. Women wanted "out" of femininity, and their outfits served as a visual sign of their determination to take a more active part in public life. Their clothing was an aesthetic stance through which they indicated to themselves and others their changing identities and modern consciousness. Through their actions, words, and appearances motoring women articulated a modern desire to fashion new versions of sexual difference. Masculine styles of dress continued after the war ended and even became high fashion for much of the 1920s, inspiring the famous "boyish" look of short hair (the shingle or the Eton crop), tweedy skirts, Oxford bags, monocles, ties, and tailored suits.

In the early 1920s images of energetic and accomplished young women were used by the automobile industry to announce the reinvigoration of a manufacturing sector that was slowly emerging from wartime production and struggling to refashion its image to meet postwar expectations of private consumption, abundance, and pleasure. "Good times ahead for the lady who drives her own car," predicted May Walker in 1919 in "A Woman's Point of View," her column in *Autocar*. "Though prices have risen enormously since 1914," she wrote, "we want to forget the strain of the last four and a half years."[7] The invocation of a female presence helped to shift automotive technology away from its recent associations with cataclysmic destructiveness. Images of attractive young women motorists, often in stylish motoring uniforms, put an optimistic spin on the future of technological progress. The November 1919 issue of *Graphic* published a full-page poster sketched by popular war illustrator Balliol Salmon of two smiling uniformed women, one in an open touring car and the other riding a motorcycle with sidecar, chauffeuring children on a country road. The poster bore the caption "A War Product: The Motoring Girl."

"Good times ahead for the lady who drives her own car." British advertisers anticipate the end of the war. *Sphere*, 5 August 1915. By permission of the National Library of Australia.

Some automobile advertisements in the interwar period referred to women's war work by using images of uniformed women drivers dressed in tunics, breeches, and high boots to sell their products. Postwar images of female chauffeurs, however, were distinctly different from the images of women transport drivers that were common during the war years. Rather than women in loosely cut, utilitarian tunics or rugged in heavy coats and boots for the front—a serious and brave response to a national crisis—images of uniformed female chauffeurs during the 1920s were more likely to be stylish and highly sexualized. The more glamorous postwar images suggested professionalism, crisp efficiency, and progressive modernity, but they also evoked the sense of a fresh beginning, excitement, and more than a hint of sexuality to the automobile industry as it struggled to meet the rising demand for private motoring. The exclusive outfitters Burberry produced a modish leather motorcycle suit in 1920 reminiscent of dispatch rider's uniforms, with a three-quarter-length coat, leather breeches, leggings, and a matching hat, and for many years after the war the large automobile accessory manufacturer Stewart-Warner used the image of a glamorous, uniformed female chauffeur, "Miss Stewart Custombuilt," to advertise its products in Britain, Australia, and the United States. Warland Dual Rim Company also used the image of a uniformed chauffeuse to publicize its system for repairing punctures.[8] The power of these sexualized images, viewed so soon after the end of the conflict, drew upon the ongoing aspirations of technologically accomplished young women, whose lives had been fundamentally changed by war.

Wearing masculine clothing was not, however, only a matter of a stylish look or fashion statement for motoring women. It flagged and augmented their claims to be taken seriously as skilled professionals—people who were legitimate participants in the working world and entitled to earn their living. Their desire for professional recognition was sanctioned and given meaning by feminist agitation for women's rights to equal employment, which had become the major focus of most factions of the British women's movement after activists suspended suffrage campaigning in favor of participation in the war effort. Motoring women were able to frame their personal aspirations, sometimes expressed in collective terms, within the wartime "right to work" campaigns formulated by patriotic feminists. The National Union of Women's Suffrage Societies (NUWSS), for example, established the Women's Service Bureau to help train and find employment for women in skilled war work. Other organizations, such as the Women's Industrial League and (after the war was over) the Women's Engineering Society, were also staunch campaigners for women's access to all forms of technical employment at equal rates of pay.

Feminist groups set up employment agencies and training courses to encourage women to gain accredited qualifications in fields of work that, unlike most munitions work, were likely to grow in demand after the war was over—be it in oxy-acetylene welding, marine engineering, auto mechanics, electrical engineering, or taxicab driving. Their publications, *Common Cause, Vote, Women's Industrial News*, and *Woman Engineer*, were filled with such schemes and highlighted that, in the words of activist and automobile enthusiast Ray Strachey, "the amazing aptitude of women for mechanical work has been one of the facts brought to light since the war."[9] Both Miss C. Griff of the Ladies' Automobile School and Gabrielle Borthwick of the Ladies Automobile Workshop predicted that the employment opportunities for women trained in technological fields would be sustained and would even increase at the end of the war.

The London branch of the NUWSS attempted to organize a union, the Society of Women Motor Drivers, similar to the Society of Women Welders, which they had established earlier in the war. All female drivers and garage employees were eligible to join, but the scheme, organizers admitted, was less than successful because privileged women were averse to unionism. The NUWSS's plans to establish a residential club and dining room in central London for women chauffeurs, however, generated greater interest. The government forced the Licensed Vehicle Workers' Union to open its membership to women in 1918, and it soon absorbed the Society of Women Motor Drivers, though professional women motorists continued to enjoy the society's residential club and restaurant in Piccadilly for another ten years.

The Honorable Miss Gabrielle Borthwick, one of the principal organizers of the Society of Women Motor Drivers and owner of the most prominent and long-lived of the women's garages, first established a motor garage in Slough and then Northwood in West London, where she taught women owner-drivers as well as trained women as chauffeurs and mechanics for business and private service. When the demand for professional training rose at the beginning of the war, she moved to central London, setting up a workshop and driving school near Hyde Park Corner and a branch in Camberwell. She claimed that hers was the only London school able to teach women to drive and maintain ambulances that had been converted to run on coal gas. Borthwick was proud that former heads of her workshop were serving in France and in Scottish Women's Hospital field hospitals in Serbia. Gabrielle Borthwick echoed Sheila O'Neil and Miss Griff's belief that motoring offered a "most suitable" occupation for women. In 1917 she declared, "I think two or three women joining together to run a garage can do very good work at the present time and there is no reason why they should not continue after the war." She continued, with a hint of frustration at her well-to-do clientele, that it was difficult, however, to make her pupils understand that "you cannot jump at once into a profession, and that there must be drudgery in learning the rudiments before attempting to take down and assemble any part of a car."[10]

Women demobilized from motor transport units continued to open new garage and chauffeuring businesses in the years immediately following the war, hoping to capitalize on the skills they had developed during their military service. In 1920 Frances Hodgson and her partner, reluctant to give up the freedom and mobility of their war work, set up a small garage, which they named the Remy Car Service. Hodgson declared that the war had provided a great impetus for women such as themselves to "strike out on lines hitherto deemed the prerogative of men."[11] It was the varied life and the independence of the work that had drawn them to that business, Hodgson wrote in a lengthy article in the *Woman Engineer*. She detected more prejudice against women drivers in civilian life than there was in the military services by the end of the war and noted that it created a problem for women's businesses, especially in London, where there was greater competition. Success, she advised prospective small garage proprietors, depended on three factors: business experience; a minimum of two cars and sufficient capital to cover running costs for at least the first year; and mechanical knowledge to help keep repair expenses down. "The great thing is to carry on, however dismal the outlook," she counseled other hopefuls. "It naturally takes time to establish a clientele, and the initial expenses are heavy, so it is wiser not to expect too great a success during the first year."

Newspapers and magazines noted similar small enterprises in those years. In 1920 a photograph in *Ladies' Pictorial* showed three women "recently demobilised from war service" who had opened a garage in Kensington. Pictured in front of a large touring car, efficient in their Women's Legion uniforms of collar and tie, belted tunic, long skirt, and high boots, Miss Ellington, Miss Mayo, and Miss Parbury, the caption said, did all the work and adjustments on the cars themselves.[12] Ivy Cummings, the well-known racing driver, ran a repair garage and used car lot on the Pultney Bridge Road throughout the 1920s. Gabrielle Borthwick's garage was listed as a Royal Automobile Club agent and a member of the Motor Trade Association. Borthwick was still providing motor services in 1928, advertising "bargains" to professional women in secondhand and small cars from the same Piccadilly premises that she had occupied during the war.

As Frances Hodgson indicated, the success of small garage and rental car businesses relied not only on qualified women wanting to take up motor work but also on the support of the men and women who patronized them. In part women's motor garages based their appeal on traditional values, seeking to cater to Victorian notions of middle-class female respectability, it being considered more "suitable" for women to be chauffeured or taught to drive by other women. But there were also distinctly forward-looking and modernist elements to their enterprise. Female chauffeurs and mechanics suggested progressive social change, and their customers, the men and women who hired them, implicitly bought into those associations. When the Independent Liberal Party candidate in the Hornsey by-election of 1921 hired a female chauffeur for his campaign, he found that he gained as much publicity for the way he moved around his electorate as for any of the policies he stood for. That publicity, however, did not always translate into permanent job opportunities for women. His chauffeur, Miss P. McOstrich, an ex-servicewoman demobilized from an ambulance corps and known as a racing driver on the Brooklands track, subsequently placed numerous classified advertisements for her services in London newspapers under the banner "Equal Opportunity for Both Sexes." The text of her advertisement hinted at women's disappointments and thwarted ambitions in those postwar years: "Ex-service woman, ambulance driver, with 24 hp landaulette, appeals for work. Contracts especially wanted; lowest terms. Drove Liberal candidate throughout Hornsey election. McOstrich, 186 Buckingham Palace Rd, Victoria."[13] Her advertisements evoke the climate of bitter disputation over women's continuing presence in the workforce, particularly in areas considered men's work, and reflected the fragility of women's place in professional automobile work.

Women's desire to remain within what had been considered masculine territory became an important element in the ethic of gender experimentation and sexual libertarianism that characterized the most daring and progressive social circles in Britain in the immediate postwar years. Public acknowledgment of an active female sexuality was becoming more commonplace, and for the first time it appeared possible that same-sex love might be considered in sympathetic terms. In that changing climate female motor drivers and mechanics provided an important imaginative resource for the lesbian networks that were tentatively emerging into public life. In one of the first literary accounts of lesbian experience in England during the first two decades of the century, *The Well of Loneliness* (1928), novelist Radclyffe Hall invoked the motorcar as an important vehicle of her heroine's sexual awakening. At the turn of the century Hall's protagonist, the aristocratic Stephen Gordon, graduated from accomplished horsewoman to competent motorist, at the same time as her inchoate sexual longings crystallized into a recognition of her same-sex desire. Owning a motorcar, which she herself was able to chauffeur and maintain, was central not only to Stephen's encounter with her first lover, Angela Crossby, but also to engineering their escape from the surveillance of Angela's husband. Stephen's car enabled the women to meet frequently, alone, and to travel to "places where lovers might sit."[14] A decade later, during the war, Stephen's work as an ambulance driver on the French front led her to Mary Llewellen, her great love, with whom she lived in Paris when the fighting was over. More than merely a new means of transport, Hall used automobiles in *The Well of Loneliness* to highlight the possibility of women's sexual autonomy from men. But it was not only in fictional accounts that early associations between automobilic independence and lesbian desire were expressed.

Garage women's air of progressive modernity in the postwar years was one expression of the fashion for female masculinities and gender-bending identities in which, as Laura Doan put it in her study of the emergence of a lesbian subculture in those years, "deviation became entangled with the chic."[15] To those "in the know," associations between sapphic desire and automobile technology added an exhilarating frisson of sexual indeterminacy to women's motor garage work. Gabrielle Borthwick, the leading female garage proprietor in London and a sister of Lord Borthwick, who founded the Theosophical Society, was a member of London's "Upper Bohemia." A spiritualist and occultist, she had been associated with lesbian literary and artistic circles since the turn of the century. For those wanting to see it, garage women and their customers represented the nucleus of an emerging lesbian identity, sometimes producing a rakish vision of sapphic love that bordered on the avant-garde in its performative dimensions.

The most notorious of these garages was named "X Garage" in joking reference to its indeterminate quality, set up in a lane off Kensington's Cromwell Gardens, by four friends demobilized from the Women's Legion Mechanical Transport Section. Standard Oil heiress Barbara "Joe" Carstairs was the principal financier, and her friends and lovers lived in a flat above the workshop, pooling their funds to buy luxurious Austin touring cars. The women put their wartime experience in Ireland and France as well as their knowledge of Italian, French, and German to good effect and advertised chauffeured holiday tours throughout Britain and the Continent. The garage specialized in taking sightseers and grieving relatives to France to view the battlefields and visit war graves. Although advertisements placed for X Garage boasted that the company's drivers would travel any distance for the lowest tariff in London, a four-day return trip to Edinburgh carried a hefty price of forty-eight pounds. A two hundred–mile weekend journey cost fourteen pounds ten shillings. Customers could store their cars and have them cared for in the X Garage workshop for twelve shillings and sixpence per week. The press published photographs of the women in overalls doing their repair work and recounted admiring stories of the celebrities they claimed as clients and their adventures on the road. The women had an arrangement with the Savoy Hotel to chauffeur guests to theaters and shows, and they were familiar figures in London's West End louche circles. Joe Carstairs, who the *Evening News* reported could dance a Charleston that few people could partner, sported the fashionable "boyish look" of Eton crop and Oxford bags and pursued affairs with various stage performers, including Tallulah Bankhead, the star of Noel Coward's *Fallen Angels* (1923) playing at the Globe Theatre.

By mid-decade, however, such privileged women's enthusiasm for motor repair work was in decline, and the public was beginning to view the ethic of gender experimentation with a great deal less tolerance. X Garage closed in 1925, and the flamboyant Joe Carstairs took up a successful career as a speedboat racer and leisured playgirl. Other women's garages, marginal businesses even in the best times, were in trouble, and those that survived beyond the mid-1920s were more modest enterprises, quietly operating without the benefit of admiring press articles or family fortunes. Borthwick Garages Ltd. was forced into receivership in 1928, but women such as Miss Ibbotson of Hampstead still posted advertisements for their chauffeured car services in the women's press. Alice Neville had closed her garage to enlist as a military driver during the war but returned to Worthing after her marriage and ran a taxicab service as Mrs. Frank Booth until 1936. Miss C. Griff, who had opened a motor garage service in London during the war, optimistically recorded in late 1925 in her aviation and motoring column in the *Woman Engineer* that the business side of women's motoring was growing in her hometown of Birmingham. She noted that a Mrs. Whitcombe had opened a garage and rental car service at Hall Green with a "nice little fleet of cars and large premises." Mrs. Whitcombe ran the business entirely herself, and, Griff claimed, she was booked up for months ahead.[16]

While feminist advocates continued to celebrate and publicize these businesses, they were in truth tiny enterprises, marginalized within an industry that was undergoing a British version of restructuring to Fordist production and vertical integration. Women's garages were undercapitalized and easily pushed to the fringes of an emerging fraternal network of automobile distributors, gas stations, and repair shops. Chauffeured rental car work made up the core of their business, but the demand decreased as automobile technology became more reliable and familiar, cars became more affordable, and people learned to drive themselves. Furthermore, the style of youthful female masculinity that had been associated with women's garages during the war years and early 1920s fell out of favor and became redefined in ways that were unsympathetic to female experimentation.

The clothing favored by garage women was similar to the military uniforms adopted by women during the war—collar and tie, tailored jackets with big pockets, sturdy boots, peaked caps, and boiler suits for repair work. But, as Laura Doan has argued, that boyish look, initially an expression of postwar high fashion through which women of all classes were able to flirt with new social freedoms without the imputation of sexual deviance, gradually came to take on more disreputable overtones.

At the height of the popularity of the masculine style the outfits, along with cropped hair and accessories such as monocles, were read as admirable expressions of modern femininity and not necessarily as markers of a lesbian identity. Heterosexual and lesbian women alike adopted the style, and masculine women were not automatically stigmatized as deviant. The joy of that sartorial toleration was precisely its playful ambiguity and the way it could confound sexual certainties. Fashionable female masculinity meant that nobody knew for certain just where any particular woman stood, creating a favorable climate for the quiet emergence of a lesbian culture. But ten years after the end of the war that tolerant climate was coming to an end. Particularly after the infamous *The Well of Loneliness* trial in 1928, in which Radcliffe Hall's novel was judged obscene and banned in Britain, such genderbending expressions of female identity suddenly appeared much less lighthearted. Photographs of Hall and her lover, Una Troubridge, dressed in a style that had previously been considered the height of modern fashion, were now published as evidence of scandalous deviance. The boyish look not only fell out of fashion, but it was unambiguously stigmatized in public forums as the "lesbian look."

That constriction in the meaning of female masculinity was a major liability for women's motor businesses. It announced the end of the admiration and tolerance for the wartime blurring of sexual difference. Female androgyny was no longer in vogue, and garage women could no longer claim to be at the forefront of stylish modernity. Images of women as accomplished and confident motorists became much less frequent in the second half of the 1920s, almost disappearing by the 1930s, and automobile manufacturers ceased to advertise their products with images of stylish female chauffeurs. The women's organizations that had previously provided the discursive and organizational resources to support women's garage work were subjected to antifeminist reaction. By the mid-1920s feminist organizations were turning away from campaigns that advocated the equality of men and women in public life, increasingly choosing to define female citizenship in less confrontational, maternalist terms. Feminist campaigning focused on women as wives and mothers, rather than skilled workers or qualified professionals. While campaigns that advocated women's equality in all forms of technical employment did not disappear altogether, by the late 1920s even the fragmentary records of professional women motorists that I have been able to assemble here are no longer evident. In her 1935 book, *Careers and Openings for Women*, feminist activist Ray Strachey, who had earlier helped to organize the Society of Women Motor Drivers, revised her wartime optimism for the future of women in technical work and noted that motor driving "is not a career which offers any prospects, and the conditions of employment are always individual."[17] This was a tremendous change from the confidence that Gladys de Havilland had expressed in *The Women's Motor Manual* of 1918, in which she had anticipated ongoing careers for women motorists in the postwar era.

Some women continued to work as professional motorists, of course, but without the public profile of earlier years, and, in the absence of a discursive climate that made their work visible and admirable, it became easy to stigmatize garage women as "unnatural," illegitimately in conflict with men over scarce jobs. By the end of the 1920s women motorists increasingly came to be represented in advertisements as wives and mothers, usually occupying only the passenger seat of the family car, rather than as competent drivers in their own right. As debate in Britain turned away from sympathetic interpretations of women's motoring ambitions, women who hoped to use their mechanical talent to earn their living had to take a backseat until the beginning of the next war, when's women's technological competence again became "speakable." Although their public success was short-lived, British motoring women's determination to push the boundaries of women's engagements with technological objects and create new ways to express a female bodily vitality constitutes an important element in twentieth-century feminist struggles to redraft the everyday, commonsense terms of masculinity and femininity.

CHAPTER 3

A Car Made by English Ladies for Others of Their Sex

The Feminist Factory and the Lady's Car

British women's ambitions to be professional motorists were formulated in terms quite different from women across the Atlantic in the United States. British women were supported by the climate of gender experimentation and female masculinity fashionable in Britain in the years surrounding World War I and drew upon the women's "right to work" campaigns orchestrated by feminist organizations at that time. They undertook a remarkable enterprise, dubbed a "feminist factory" by the press, in which middle-class British women were involved in manufacturing an automobile that was launched onto the British market in 1920. Called the Galloway, after the region of Scotland in which it was produced, it was described by one contemporary commentator as "a car made by ladies for others of their sex."[1] Like Hilda Ward some ten years earlier in the United States, the women who worked at Galloway Motors Ltd. aspired to hold tools in their hands, though the tools they desired to use were not the wrenches and screwdrivers of home repair but the large machine tools of industrial production.

The first "pleasure cars" produced in Britain after the end of the war attracted a great deal of interest. Years of wartime restrictions and enforced savings had created a strong demand for consumer goods, and the motoring and general press debated at length the future of the British auto industry. Given that the United States had emerged from the war with a greatly strengthened manufacturing sector, there were questions about how the British industry should respond to the threat of American mass production. There was also a great deal of debate and industrial conflict over just which changes in industrial relations introduced under wartime conditions would be carried over into peacetime. Commentators noted alterations in women's relationship to automobile consumption, observing that women preferred to drive themselves since the war and often maintained and serviced their cars as well. Reporting on London's first Motor Show after the war, one female journalist wrote that a significant proportion of the visitors were female drivers in motoring dress. Manufacturers' representatives were "astonished" by the amount of technical knowledge shown by lady clients who wanted to know how the "various working parts could be reached and kept in good running order."[2] Manufacturers had paid great attention to making components accessible and simplifying maintenance, "an accomplishment in which many ladies are quite expert," the journalist declared.

The Galloway car was one response to the challenges of that renewed consumer demand immediately after the war. It was manufactured by the noted Scottish automobile company, Arrol-Johnston, in an ultramodern factory in the rural village of Tongland in southwestern Scotland. *Queen, the Ladies' Newspaper and Court Chronicle* suggested that the "interesting new car" might be especially attractive to its readers because the car was produced in a factory run by women operatives during the war.[3] The Galloway crystallized a particular historical moment in which a "ladies' car" could be a material expression of a broad range of technological and social changes in the years surrounding World War I. Philosophies of industrial relations, class struggles over workshop control, the application of Fordist methods of mass production, technical advances derived from military hardware, ideas about the future of private motoring, and contestation over changes in gender relations came together in the Galloway car—so much so that the Galloway has come to represent a material expression of a particular historical moment in British women's changing relationship to commodity consumption, technical expertise, and industrial production.

The Galloway was a modest two-seater coupe with a rear fold-out dickie seat—a layout that was considered the lady driver's car par excellence. The reviews published on its launch were favorable, with journalists declaring that it featured "well thought-out arrangements" and had "many unusual features incorporated in its design," which would be of particular interest to women drivers. They noted with approval that the components were well placed and accessible, making for easy maintenance and repair. The car was light and simple to drive, with a smooth gear change action. The coachwork was of superior quality and the interior design roomy, with "sumptuous upholstery." Unlike some other light cars, it had two doors so that the driver did not have to clamber over the passenger seat, and the driver's seat was set high to improve visibility over hedges. The rake of the steering wheel could be altered to suit the driver, and the brake and clutch pedals were also adjustable. It had disc wheels, which were easier to clean than the spoked artillery wheels more commonly used in British cars, giving the Galloway something of the stylish, racy look of Continental cars.

"The whole car runs with a most desirable 'one-piece' feeling, rare indeed in its price-class, and none too common higher up the scale," one reviewer enthused. It had "an astoundingly competent little engine" that did its work "with an entire absence of fuss or noise," "getting along with mere whiffs of fuel." The car's links with wartime aero-engine design were carefully noted, and the use of aluminum alloys in gear wheels to reduce wear and noise and in other components for weight reduction were postwar innovations that generated great interest. Ominously, given the highly competitive climate of postwar British automobile manufacturing, one prominent reviewer concluded, "Everywhere about the Galloway are evidences of the fact that [works manager and chief engineer] T. C. Pullinger settled the design and material and workmanship of his car long before he bothered himself about its price."[4]

Promotional material for the Galloway emphasized that the car was manufactured in a factory of modern design in the "idyllic braes and glens" of rural Scotland, rather than in the notorious heavy industrial region of the Clyde Valley. By setting the plant in a greenfield site and by taking advantage of natural lighting and cheap turbine power "stolen" from mountain streams, the company claimed it was able to produce cars in which overhead costs and production delays would be kept to a minimum. At a time when the engineering sector faced massive confrontations between management and trade unions over workshop control, the automobile columnist for *Queen* endorsed the wisdom of locating the Galloway factory in a district "not invaded by the labour agitator." Veteran motoring commentator in the *Graphic*, Edwin Campbell, praised the Galloway Engineering Company as an "experiment in social economy." Its cooperative structure, in which the capital was held entirely by the directors, production workers, office staff, and sales agents, ensured that "every person concerned with its production and sale will have every incentive to make the project a success."[5] The company was set up to secure the maximum economy in production costs and to eliminate disruption through strikes, he declared, and other motor manufacturers would keenly watch its progress.

The site of this experimental factory was the tiny rural village of Tongland in southwestern Scotland, two miles from the ancient fishing port of Kirkcudbright. Unlike most munitions factories during World War I, in which existing engineering works were diverted into wartime production to fill a short-term need, the Tongland plant was purpose-built with the future in mind. Its very appearance, an impressive four-story factory that used the latest techniques of industrial architecture developed in Detroit automobile plants, was designed to signal and shape its special purpose. But it was not just the ultramodern plant in a remote rural setting that made the factory remarkable. The enterprise offered the promise of a bright future for middle-class women in industrial production. Some press articles even called it "an engineering college for ladies" and "a fine University for Women Engineers," rather than merely a factory. *Lady's Pictorial* declared it "a paradise for the woman engineer."[6] For women who aspired to become professional engineers, the factory building was a declaration in glass and ferro-concrete of permanent social change.

Galloway Motors was a subsidiary of Arrol-Johnston, one of the oldest and largest of the Scottish car manufacturers of the Edwardian era. Arrol-Johnston's managing director, Thomas Charles Pullinger, had joined the company five years before the war, and one of his first actions was to move the company's production out of the heavy industrial region of the Clyde Valley, with its labor militancy, into a plant he had specially built on a rural site at Heathhall, near Dumfries. When war was declared, the Heathhall factory was converted from automobile to aircraft production under government contract. With his daughter, Dorothée Pullinger, who had started her engineering training in the Arrol-Johnston drawing office before the war, T. C. Pullinger conceived of a subsidiary, the Galloway Engineering Company, that would train and employ women engineers to supply aero-engine components to the Heathhall plant twenty miles away.

Like Arrol-Johnston's larger plant at Heathhall, the Tongland factory was built following designs patented by the leading industrial architecture firm in the United States, Albert Kahn Associates. It was reminiscent of Ford's Highland Park works, in Detroit, though on a much smaller scale and without the assembly line production or rational flow of materials from the upper stories to the ground floor. The building was constructed around a steel reinforced concrete frame that allowed for full-length glass windows on each floor. Heralded as a "daylight" factory, it reflected the new emphasis on the welfare of industrial workers in Britain, as women entered the industrial workforce in growing numbers. Offices and assembling workshops were placed on the first three floors, with the machine shop and foundry housed in an adjoining building for fire protection. The fourth floor was a recreation area that included an "all-electric" kitchen and dining room, library, and games room. There was a tennis court on the flat roof and a swimming pool constructed into the river. At a time when local towns were still twenty years away from electrification, the factory produced its own hydroelectricity, with turbines powered by the River Dee. The building has survived, though it is now in a derelict state.

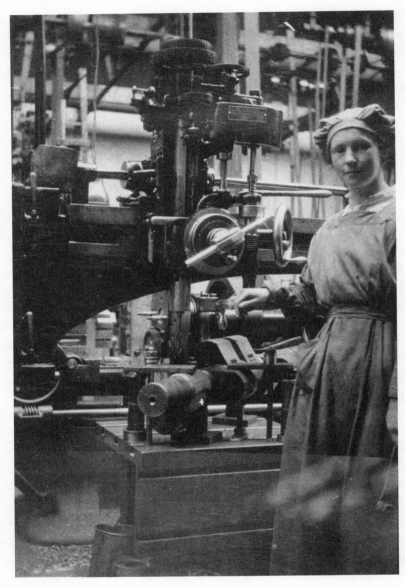

"A paradise for the woman engineer." Boring engine connecting rod forgings at the Tongland Factory, Scotland. *Engineering*, 9 November 1917. By permission of the National Library of Australia.

The Tongland factory opened in early 1917 with a core of ten women who had had four months' special training at the Glasgow Institute of Technology. They cleared building rubble in the bitter cold of the infamous "freezing winter" of the war, cleaned rust from machine tools, and set up the workshop under conditions they described as "second best to Flanders." Nine months later there were some sixty women working under two male engineering instructors and two female supervisors. In an effort to attract more of the "right kind of woman" to the district, the company organized press tours of the factory and produced a glossy, illustrated brochure. It expressed confidence that there were sufficient "educated gentlewomen" wishing to adopt engineering as a permanent profession, rather than choose the mere "cul-de-sac" of employment in shell factories for the duration of the war.[7]

Engineering journals and magazines targeted to upper- and middle-class English readers published glowing reports on the factory "somewhere in Scotland." *Gentlewoman* assured its readers: "Wonderful was the cleanliness and orderliness of the factory—no dirt, no oil reek, no piles of rubbish. Warmed by electric radiators and perfectly ventilated, it seemed an ideal 'shop' indeed for women engineers."[8] In his enthusiasm for the project managing director T. C. Pullinger painted an extravagant picture of what an "ambitious gentlewoman" might hope to achieve in the engineering profession, promising high wages and declaring that "the trained woman engineer has come to stay."

Promotional brochures for the Galloway Engineering Company emphasized the beauty of the area and suggested that employees would be able to ramble through romantic countryside in their free time, paint landscapes, write poetry, and study botany. Pullinger was convinced that the site's nonindustrial setting would appeal to privileged young women, though he later acknowledged that its isolation was a drawback that tested the enthusiasm of even the most willing "girl engineer." The company provided accommodations for a small number of women in the village of Tongland, but, with more than two hundred workers employed at its height, most had to board two miles away in the small fishing town of Kirkcudbright, privately or in one of several company hostels.

Kirkcudbright's population was an eclectic mix even before the factory was built. "The most artistic town in the United Kingdom," it had been called, having become an artists' colony favored by members of the radical Glasgow School of Art, including some of the feminist "Glasgow Girls" attracted to the Green Gate Close studio of Jessie M. King. The Glasgow School was characterized by its focus on industrial design and its close connections with industry. Tongland workers formed friendships with the artists, and together they played an active part in the life of the township, organizing entertainments, parades, and fund-raising lectures. Vera

Holmes, who had been a chauffeur to suffrage leader Emmeline Pankhurst before the war, lived in Kirkcudbright for a time. She had served as an ambulance driver with the Scottish Women's Hospitals in Serbia, and the Tongland women organized fund-raising benefits in the Kirkcudbright town hall at which Miss Holmes showed lantern slides of the Scottish Women's Hospital's Serbian work.

Working conditions at Tongland reflected the somewhat awkward mix of the enterprise's educational and industrial purpose. Unlike the brief training programs offered to munitions workers elsewhere, in which women were given practical instruction in just one task, compulsory theory classes during working hours and lessons in drawing and drafting were planned as an integral part of the work at Tongland. The women were expected to sign up for a three-year apprenticeship, in which the rates of pay were higher than for normal male apprenticeships but considerably less than the adult wage paid to other munitions workers.[9] Starting wages did not cover the cost of food and accommodations or the cost of relocating to that remote area. Records of the Women's Engineering Society (WES), which was formed immediately after the war to support prospective women engineers, show that most workers were young and single and had traveled from England to take up the work, though some came from the surrounding district.

No personal accounts have so far emerged of how the women experienced their time at the factory and the best indication to date can be found in the four editions of their works magazine, the *Limit*, which were preserved in a Kirkcudbright art gallery. The name they chose for their magazine is revealing and hints at some of the hopes and ambitions the women invested in their work. *Limit* is an engineering term for "a boundary restricting measurement"—that is, the limit of accuracy—but the term suggested many things besides. Women's capacity for accurate measurement had been ridiculed in the press in the early days of their movement into munitions work, and the Tongland women's choice of name suggests that they had embraced the controversy with a sense of irony. They were defying restrictions, pushing the limitations of femininity, and making a humorous reference to their audacity: "Women engineers—that's the limit!" Beyond that, the name hinted at the way in which women's engineering work was subjected to unprecedented scrutiny in those war years. Journalists and popular writers with little industrial knowledge emerged from visits to munitions factories hugely impressed with what they saw and produced a great deal of extravagant prose about frail women bravely "taming mechanical monsters" for the war effort.

The *Limit* provided both workers and management at the Galloway Engineering

Company with a forum for exploring their experience of working under wartime conditions. Several articles reveal the women's sense of being "on trial," of being closely observed by a skeptical and sometimes hostile audience. They were conscious that, through their pleasure in the new knowledges and competencies they were acquiring, they were creating a fresh female identity not just for themselves but also for the benefit of a broader, national audience. On the first anniversary of the factory a woman who signed herself "Pioneer" wrote about their awareness of being on display as the "first girl engineer apprentices," out to prove that "girls can be boys, as far as mechanics are concerned." A correspondent for *Autocar* agreed. These patriot machine hands were showing how "we are breeding a new race of girls," he declared.[10]

Flight was the glamour technology of the day, and the work at the Tongland factory in aero-engine construction was at the high end of skilled manufacturing. The women were involved in establishing an advanced new industry, which they believed was appropriately staffed by an advanced new workforce. As they machined lightweight, high-revving engine components in aluminum and new alloy steels to fine limits, with a degree of accuracy and standardization new to British manufacturing, they hoped that would secure more openings for women in engineering. Those feminist overtones to the enterprise, however, were rarely spelled out. Most women who worked at Tongland were some years younger than the prewar suffrage activists and were careful not to have their actions identified in terms of a war between men and women, as the suffrage movement had been characterized. Knowing that their work was releasing men to fight in the trenches and providing them with matériel made it quite unthinkable to sustain notions of a "sex war." Besides, the women were daily dependent on the goodwill of male instructors. Accordingly, like others of their generation, they were inclined to disavow or distance themselves from overtly feminist affiliations and proceed as if economic and professional equality between the sexes had already arrived. They would quietly enact and demonstrate their competence, rather than stridently demand equality.

There were, however, symbolic and personal links with earlier feminist campaigns, particularly through the involvement of older women such as Dorothy Rowbotham, a mathematics tripos from Girton College. Those links were given little public emphasis by the company, though some sections of the press called the Tongland plant "the Feminist Munition Factory."[11] Yet the signs were there to those who were sensitive to them. The colors of the largest suffrage organization, the National Union of Women's Suffrage Societies—green, red, and white—were used in the cloth badge the women wore on their overalls, and after the war enameled plaques distributed to sales agents for Galloway cars featured the purple, white, and green of the Women's Social and Political Union.

While the women tended to distance themselves from overtly feminist claims, they were enmeshed within a history of class and gender conflict in industrial production from which they could not easily extract themselves. The Tongland factory was a "Government Controlled Establishment" under the Ministry of Munitions, a new government department established to oversee wartime production and to broker a truce between male craft unions and industrialists. Battles over craftsmen's control of workshops had been waged between the engineering unions and manufacturers since the 1890s, when British industry came under increasing challenge from German and American competition. Government intervention through the Ministry of Munitions in the early years of the war forced an agreement so that what had been skilled jobs could be broken down into skilled and unskilled components. The change allowed an inexperienced operator to learn to perform a single process quickly on a machine that had been set up for her by a skilled operator, who then controlled numerous machines.

"Dilution" was what this wartime solution was called, and both sides, labor and management, firmly believed that the other had gained the best of the bargain. It benefited those men who remained in the industry because it upgraded their skills and turned them into supervisors, but it also signaled postwar changes to work practices. Male craft workers were rightly concerned that employers would fight to continue employing female labor and attempt to use wartime changes in the definitions of skill, pay rates, and workplace practices against them after the war ended. Unions feared that dilution would permanently reduce the number of skilled men when the war ended and would expand the use of unskilled labor, leading to work intensification. Employers, for their part, were angered by government moves to draw unions into the planning of wartime production and indignant that they were not left to manage their factories as they saw fit.

Scottish manufacturers were known for their trenchant antiunionism and confrontational approach to industrial relations. Unionists had been excluded from many engineering workshops before the war, and prominent Clydeside industrialists deliberately precipitated industrial confrontation in 1915 and early 1916, to the consternation of Ministry of Munitions officials, who were not yet in a position to impose the new Munitions Act in the face of mass action on the part of workers. By the middle of 1916, however, Scottish industrialists were forced into compromise by the pressure of wartime contracts and patriotic sentiments—except, that is, for T. C. Pullinger, who had committed the Galloway Engineering Company to another course of action. His Tongland factory represented a move against the pressure for industrial change epitomized by the assembly line, even though it was equipped with some of the latest machine tools then available in Britain and in spite of the building's Fordist echoes. Paradoxically, rather than being used to break down the broad base of traditional craft skill that wartime dilution implied, the women at the Tongland factory were being inducted into it. They were taught to read drawings, use micrometers, and set up and maintain their own machines and were instructed in tool making and the manufacture of jigs — the templates needed to adapt generic machine tools to specialized repetition work.

T. C. Pullinger had played an important role in introducing American production methods into the British automobile industry in the prewar years. In his previous position as manager at Humber's Beetson plant he had been locked in bitter disputes over workshop control with the Amalgamated Society of Engineers (ASE), the exclusively male craft union that covered the majority of automobile workers. When Pullinger took over the management of Arrol-Johnston Company, part of his brief was to defeat the unions. His decision to move the company out of the militant Clyde Valley into an area with no strong engineering tradition was a deliberate attempt to exclude unions from his factory. During the war Ministry of Munitions policy obliged Pullinger to accept union organization, but, as soon as the war was over, he again set about ejecting ASE members from his works.

Yet Pullinger's strong antiunion sentiment was more complex than it first appears, for he did not simply link a nonunionized workforce with the direct imposition of a Fordist system of production, as was the case in the United States. As a reviewer of the Galloway car suggested in 1920, while noting that Pullinger was more interested in the design and workmanship of his products than in the economics of their production and distribution, T. C. Pullinger was a designer and mechanical engineer before he was a businessman. His interests lay in technical perfection and quality of output, rather than in mass production. As a deeply religious man, conscious of the mechanized slaughter on the battlefields in France, he sought an alternative to an assembly line that placed workers subordinate to machines. In 1917, when the Tongland works was coming into production, Pullinger declared, "Our operatives must have instilled into them the idea that their work is an art, and that it is a high privilege to be able to operate machine tools, and produce beautifully finished interchangeable parts."[12]

That vision of industrial relations, in which he characterized manual labor as a creative activity on a par with sculpting or playing a musical instrument, hinted at nostalgia for an era of daily face-to-face relations within autonomous factory communities, such as the model village founded in Bournville in the nineteenth century by the wealthy Quaker businessmen, the Cadbury brothers. Pullinger evoked an imagined past of commonality of interest between labor and management, in which managers had only benign goodwill for workers and workers willingly accepted managerial authority. His experiment with an all-female workshop in a closed community was one expression of that belief, but it was a new, modernized version of that agreeable workplace, forward-looking and backward-looking at the same time. From the company's point of view, it had several distinct advantages. Government policy and public funding supported the factory and provided subsidies for its machinery. By employing women, it was tapping into the ultimate greenfields labor force and sidestepped the power of the unions. Its female workforce was likely to prove compliant and grateful for the opportunity, as women such as his daughter, Dorothée, and other founding members of the Women's Engineering Society had been pressing for admission into engineering work for some time.

Pullinger confessed to being a reluctant convert to the idea of women engineers, but he soon expressed great satisfaction with his experiment in industrial relations, describing its success in terms that were to become standard for characterizing women's industrial skill. Women's sense of touch was more accurate than men's, he often proclaimed. They broke fewer drill bits, were particularly suited to inspection work, and were much more amenable than men to finicky, repetitive work. "They are born mechanics, who work with their brains as well as their hands," he claimed. "I have found that a woman's touch is more trustworthy than a man's. She seems to have a special instinct. I am convinced that there is an immense future in engineering for women who really love their work and are keen on it."[13]

Pullinger's conversion to the value of women engineers, however, was clearly undergone with an eye to their value in postwar labor relations. In his opening address as president to the Institution of Automobile Engineers in 1917, he raised the question of whether the trade unions would permit manufacturers to continue using female labor after the war. Nobody could be sure that women would be able to remain, but manufacturing companies needed to position themselves to take advantage of the possibility, he warned: "My own opinion is that it would be a grave mistake to eliminate the asset of female labour in helping us to fight our industrial battles in post-war times."

For a time the women at Tongland seemed to experience their work as an art and high privilege, as Pullinger had hoped. Their pleasure in the work, its new vocabulary, and the physical skills they acquired are reflected in the pages of their magazine *Limit*. Even during the industrial turmoil and uncertainty of the immediate postwar period, and despite the hostility expressed toward them by unemployed veterans and unionists, many stayed with the enterprise. They did short-term work building tractor and truck motors for outside contractors while they waited for the Tongland works to reorganize and develop its own postwar product. But it became increasingly obvious that, even if the company did survive the postwar transition, it would not be in the form of an idealized "fine university for women engineers" that the women had earlier been promised. For, in spite of some successes, the Galloway Engineering Company did not prosper.

There were design and production faults with Arrol-Johnston's aero-engines, and other manufacturers were able to snare the best of the contracts. In addition, a middle-class female workforce did not come forward in the numbers the company had projected. Geographical isolation, poor wages, and lack of housing all played a part, and the need to keep the factory's machines working at full capacity meant that educational objectives came to take second place. As the pressure of production schedules began to dominate the life of the factory, the company began recruiting working-class women—local and Irish women of the "industrial type," as one report put it—rather than the preferred middle-class trainee engineers from the south. Lectures in engineering became voluntary and were increasingly held outside of working hours, and students were required to pay a fee for their training. By the end of the war Dorothy Rowbotham, who had been the female supervisor from the outset, estimated that of the two hundred workers only about fifty followed the training program. She confided with palpable disappointment in an interview with the president of the National Council for Women, "The force of war conditions has gradually turned the college into a works." About the future of the business she was pessimistic: "It remains to be seen whether the educated woman can stand the monotony over the necessary number of years."[14]

Rowbotham's assessment implicitly recognized the profound class contradictions of the enterprise, which became ever more apparent as the end of the war removed justifications for the higher value of the work. Far from being glamorous engineering, as promised in brochures and magazine articles, most of the activity was repetitive machining, and no amount of pride in the work, bodily pleasure in new skills, or pioneering zeal could hide the reality of its nonprofessional status. For all her upbeat public statements "that women can set up the most complicated machines," Rowbotham berated employees in the pages of the *Limit* for not having achieved the same productivity as men and women in the shell factories. While the women had developed a high level of proficiency in a variety of tasks, she pointed out, their output was not equal to that of skilled workers who performed just one task.

By 1920 the optimistic statements in the first editions of the works' newspaper had turned to disillusionment and savage humor against the false promises of professional training and skilled work. In a parody of some of the early buoyant pronouncements on the future of women in engineering, a worker signing herself as "S. Suds," in reference to the soapy water used to cool and lubricate metal during the machining process, wrote:

> There is a certain factory in Scotland (oh, famous phrase) where girls are generously allowed to train in all branches of Labouring and Scavenging. They are allowed, quite openly, to clean their machines, remove the chips, fill their suds tanks, and to fetch and carry the work to and from their machines . . .
>
> Very little more need be said to assure the women of today of the desirability of entering such an attractive profession, and to urge the educated and refined gentlewoman to enter the course of training at once . . . Knowledge of Hebrew and Botany will be found invaluable.[15]

Many contemporary commentators noted just how quickly the discourse celebrating female technical competence turned against women when the war was over. As S. Suds's bitter assessment reveals, however, it was not only those who were hostile to women's industrial work but some of the women themselves who came to revise the value of their work—and the pressures to do so were enormous. Male unions were agitating for a return to prewar conditions in manufacturing, and ex-servicemen's organizations were demanding the preference in postwar employment that they had been promised in the early years of the war. The postwar climate of wage cuts, prolonged strikes, ex-servicemen's militancy, inflation, housing shortages, and the Restoration of Pre-War Practices Act were felt in the most personal terms in the small town of Kirkcudbright. Demobilized servicemen came home to find a brand-new factory in their district, staffed by educated "girl engineers," but little work for them. While they had been away practicing the skills of war, privileged women, and strangers at that, had moved into their town to acquire the skills of the future. The men's dissatisfaction surfaced during the planning of the Kirkcudbright Peace Celebration parade in mid-1919. Only a month earlier the *Kirkcudbright Advertiser* had published a feature article on the success of the Tongland factory, predicting that women engineers would play an important part in postwar reconstruction, but the local chapter of the Comrades of the Great War, with a membership of 120 and led by the local priest, was less than ready to celebrate—either the peace or the part women aspired to play in it.

The local chapter greatly alarmed the township by threatening not to turn out for the Peace parade. The focus of its members' discontent was the Tongland factory, which they characterized as "a place run for women with a sprinkling of men."

They objected to the "continued employment of women in industries formerly open to men only" and "the Tongland girls, who were nearly all strangers." The veterans were eventually persuaded to march at the head of the parade, with the Tongland floats placed at the rear, but later they presented the town council with a petition demanding that work be found for them. The town council responded, using the same defense that the company itself employed. The works had been specially built for women, they argued, and it had been distinctly stated that the jobs were to be for women and women only. Besides, "the girls were not displacing any of them, none of them having been employed there before the war had started."[16]

The argument did not placate the men, and discussions became even more acrimonious, hinting at the bitter divisions in the town. Councilors suggested that many of the men were not suitable to employers, declaring that some of them did not want to work and were even "refusing work point blank." One created uproar by claiming that, far from being heroes, the men "went to war at the point of a bayonet" and so did not deserve special consideration.[17] The debate continued in the letters column for a time but eventually faded from view after a committee was appointed to recommend the employment of ex-servicemen to employers, "without going to extremes," but pressure remained on the women at the Tongland factory.

The Women's Engineering Society, which had been formed in 1919 to promote women's continued presence in engineering, closely monitored developments in Tongland. To provide moral support for women who remained at Tongland, WES representatives visited the factory and set up a Kirkcudbright branch in early 1919.[18] They signed up ninety-six of the workers as associate members. For a time it was an active branch, holding social events and professional activities. The Kirkcudbright branch sent delegates to WES meetings in London and scheduled lectures on topics such as "The Tractor Engine and Its Testing," "The Manufacture of the Magneto," and "The Position of Women—Past, Present and Future." The engineering society fed the women's press optimistic stories about Galloway Motors and referred unemployed women with engineering ambitions to the Tongland works. Privately, however, they discussed the difficulties that the company was facing and devised strategies to keep that symbol of women's future in the engineering trades viable. They sought advice from sympathetic industrialists on what kind of work the factory might take up, considered various investment schemes to rescue the enterprise, and sought the patronage of wealthy sympathizers.

The WES tried to straddle the class divide between skilled craftwork and professional engineering throughout the industrial relations turmoil of 1919 and 1920. The group attempted to negotiate with the Amalgamated Engineering Union (AEU), which had replaced the Amalgamated Society of Engineers, on admitting women, but the union refused to meet with them. The AEU was fighting for its life in the changed climate of industrial unionism and had organized many of its campaigns around preventing the extension of female labor. By 1922 the WES had given up on a cross-class agenda, concentrating exclusively on women's admission to professional engineering associations, where they were having a little more success.[19] Those difficulties were reflected in their membership rolls. By 1920 only forty-one Tongland women remained paid-up WES members, and by 1921 the number was down to twenty-four, with most indicating the uncertainty of their future by paying quarterly subscriptions.

T. C. Pullinger's first postwar priority had been to see Arrol-Johnston's large Heathhall factory back into car production, but by 1920 he was able to turn his attention to the Tongland plant. When he announced the Galloway car project, it brought a rush of optimism to the Tongland works and to the Women's Engineering Society in London. Three two-seater Galloways were hastily built for the all-important London motor show at Olympia in November 1920, and early press reports on the car's launch were encouraging. It was economical, easy to maintain, well finished, and had a solid "hand-built" feel, they said, at a time when standardization and mass production were increasingly the order of the day. Its price, however, was a major problem, as the buoyant demand and high prices paid for cars immediately after the end of the war had slumped by 1921. The two-seater coupé was advertised at 550 pounds in 1920, while a comparable Ford touring car, even after prohibitive import taxes, sold at 275 pounds. Even though the price of the Galloway had been halved by 1923, it was still 100 pounds more than a similar mass-produced British-built Morris.

Thus, the Galloway was not to be the solution for the women at Tongland, and they steadily left the plant. Even works supervisor Dorothy Rowbotham announced that she would leave to start up her own engineering business in July of 1920, though she stayed on when plans to manufacture the Galloway car were revealed. A small number of cars, perhaps no more than one or two hundred, were produced at the Tongland works until March 1922, when the factory was closed and production transferred to the Heathhall factory, with its male labor force. Like Arrol-Johnston's other models, the marque turned little profit, and the company staggered on until 1929, when production of the Arrol-Johnston ceased altogether. In all, probably less than four thousand Galloways were manufactured, but the company's problems were by no means unique. What had been a healthy and proudly craft-based prewar Scottish motor industry was well and truly finished by the end of the 1920s. Pullinger's choice of industrial organization and his preference for women employees in automobile production had proven not to be the way of the future. His cars did not survive the economic recession that coincided with the Galloway's launch, nor could they compete with the lower prices and more exciting design of English and American models built on assembly line principles.

As with much research into women's histories, there remain many gaps in the story of the Tongland factory. Questions cry out to be answered and perhaps will always remain that way. Dominating the pages of the *Limit*, as well as the perception of the Kirkcudbright Comrades of the Great War, were the middle-class English women, especially the first cohort of workers, "the pioneers." Broad hints of their sense of class and imperial superiority are found in the ways they defined themselves against the "quaint" Scots townsfolk and the Irish "Colleens" and "Biddies" who came to work in the factory in later years. Occasionally, the *Limit* published muted protests from another group who called themselves "the pygmies," but the voices of other workers and knowledge of the subsequent course of their lives may never emerge. What is clear, however, is that it was not the middle-class women seeking professional qualifications who remained within the factory system in the postwar years. Rather, it was working-class women of the "industrial type" who would continue to work in it—even coming to dominate the new manufacturing sectors that produced consumer durables. The fears of the male craft unions proved to be well founded, as the wartime introduction of women into manufacturing went hand in hand with the long-term changes in work practices that employers had been trying to introduce for many years. Women's move into industrial production in the immediate postwar years provided the basis for a new division of labor in which working-class women and men were characterized in terms of "natural" differences, with women excelling in (low-paid) repetitive tasks and the earlier "aristocrats of labor," skilled craftsmen, indeed finding themselves to be the losers of modernity.

As for the middle-class girl engineers at the Galloway Engineering Company, they continued to negotiate their way through the opportunities and limitations they encountered in the postwar climate. Some continued to find employment in small engineering works. Dorothy Rowbotham and Miss Lees found work at Margaret Partridge and Company, Electrical Engineers, which specialized in farm and domestic electricity supply. Miss Bridge became the organizer for a new branch of the WES in Bedford; Miss E. F. Bull studied for a bachelor of applied science degree (in metallurgy) at the Glasgow Technical College; and Dora Turner and Annette Ashberry studied for a bachelor of science degree (in engineering) at Loughborough Technical College. When six of the most skilled and enthusiastic of the Tongland women left in October 1921 to set up their own small cooperative engineering works in the Midlands, the WES threw its energy behind them.[20] It promoted their company, Atalanta Ltd., in the *Woman Engineer* and in other feminist papers, placed prominent members of the WES on the board of directors, and announced women's design competitions for inventions that the Atalanta factory might patent and manufacture.

One woman who remained involved with Galloway Motors was Dorothée Pullinger. She returned to Arrol-Johnston after her wartime position as female supervisor at the giant Vickers munitions works at Barrow-in-Furness ended. In 1919 she set about learning all aspects of the business, beginning in the foundry of the Tongland factory, where she learned core making, molding, and casting. It was a combative, antiunion stance, because there were acrimonious national molders' strikes at the time, and one of the conditions for the men's return to work was that women be "let go." After the Tongland works closed and production was transferred to the Heathhall factory, Dorothée Pullinger remained with the company another four years, though not as an engineer but as a sales representative for southern England. She was a successful competitive driver of Galloway cars in trials and hill climbs and remained an active councilor of the WES and member until she died in 1986.

Dorothée Pullinger best exemplifies the creative approach adopted by women engineers in their search for meaningful work during those postwar years. Late in her life, in response to an inquiry from a researcher, she wrote, "I am a member of the Chartered Institute of Professional Engineers and also have worked most of the time in Engineering, till I was told I was doing a man out of a job, so I went washing."[21] It was a tongue-in-cheek reply that hints at a joke shared between women engineers of her generation—a joke about both the class and the sexual division of labor in postwar Britain. Immediately after the war, sections of the press elevated laundry work and domestic service into markers of the normalization of class and gender relations. They depicted the unwillingness of women munitions workers to return to those prewar forms of women's work as symptomatic of the ways that war had upset natural class and sex hierarchies. Laundry work, low paid and arduous as it is, had long been a safety net for women in times of adversity. It was a "woman's skill" that could be adapted to domestic responsibilities and a shortage of capital more readily than most occupations. Pullinger's own grandmother had opened a hand laundry in Great Portland Street to support herself and her four children after her husband's death.

Dorothée Pullinger, however, did not just "take in washing," as her joke suggests, and she declared that it was a man from the Laundry Machinery Company who taught her how to wash. As a middle-class woman with technical expertise and access to capital, she was able to approach washing as an engineering enterprise. She imported the latest automatic steam laundry equipment from the United States, and her White Service Steam Laundry of Croydon opened just in time to catch a boom in the commercial laundry industry. For Dorothée Pullinger her laundry operation was an engineering enterprise disguised as "woman's work." Nobody could object to a woman who went washing. As a trespasser into the male domain of professional engineering, Pullinger and other women like her in those interwar years were forced to create quiet, personal solutions to a social impasse.

The Galloway Engineering Company was founded on women's desires to participate in larger processes of political, technological, industrial, and economic change in the years surrounding World War I. A particularly British expression of female modernity, the enterprise provides an alternate view of the struggle between labor and capital in those years and highlights how feminist action could be both progressive and reactionary in the same moment. The forgotten story of these aspiring engineers is part of a larger history in which women of each generation have been obliged to reinvent, as if for the first time, new relations of female technical competence. Most important, the history of this short-lived enterprise calls us to be sensitive to the ways that, by aspiring to machine aero-engines and automobiles, these women were also machining new versions of the category "woman." In their words and actions they were calling into question notions of a fixed interval that separated men and women, thereby trying to formulate ways by which they too might move differently in the postwar era.

CHAPTER 4

Transcontinental Travel

*The Politics of Automobile Consumption
in the United States*

Women's patriotism in the face of national crisis and the feminist movement's campaigns to secure women's access to careers on equal terms with men provided a political climate for women's personal and professional ambitions in the years surrounding World War I. Following the war, the continuing class exclusivity of motoring and the fashion for female masculinity, or the "boyish look," lent an aura of modernity and glamour to uniformed women chauffeurs and mechanics in Britain. In the United States, however, women's enthusiasm for automobiles in those early decades was being differently expressed, in both public forums and by women motorists themselves.

The war did not bring the same degree of personal trauma and social dislocation in the United States as it had in Britain, nor did it precipitate a similar sense of crisis in the meanings of gender difference. Instead, it marked the creation of the first mass consumer culture and an upsurge of social optimism that lasted until the late 1920s. The war had been a turning point in global power, with Europe in decline and the United States confirmed as the world's largest economy. For a broad section of the American population the postwar era promised unheard-of levels of prosperity and provided a much more rapid diffusion of automobiles across social classes, compared to Britain. Car ownership became emblematic of an energetic and particularly American democratic national culture.

In 1923 *Motor* magazine could proudly declare that New York State alone had more automobiles than any foreign country and that nine-tenths of the world's cars were owned in the United States of America. There were about seven million passenger cars in the United States in 1919, but by 1929 the number was almost five times greater—even more than the number of households with telephones.[1] While people of color and recent immigrants were certainly buying cars, it was predominantly a technology of nativist, white privilege. Fears about the effects of rapid urbanization, with the increasing heterogeneity of cities through migration from both Eastern Europe and the movement of African Americans from the South, led manufacturers and their agents to emphasize the automobile as a vehicle that could return white, middle-class Americans to their roots in the rural landscape through leisured automobile touring.

Along with high levels of economic prosperity, the early decades of the twentieth century also saw a public struggle for women to be recognized as citizens with a stake in the life of the nation, so that commodity consumption and female citizenship became closely entwined. Much more than a purely economic activity, commodity consumption became central to women's identities as active citizens in the world beyond the home, thereby acquiring political dimensions. In promoting mass-produced products, manufacturers advanced an ethic of individual fulfillment, personal choice, and female independence that meshed closely with the demands and aspirations of the suffrage movement. That conjunction of feminism and commodity consumption was expressed much more forcefully in the United States than in Britain, and the terms within which it was framed were quite different.

The immediate discursive climate for independent and energetic femininity in the United States was provided not so much by the political aspirations of the feminist movement, as it had been in Britain, but primarily by images of modern femininity emanating from Hollywood. While accomplished women motorists were sometimes represented in overtly feminist terms in the United States, it was a feminism much less inclined to question the borders of masculinity and femininity through performances of female masculinity than that in Britain. On the other side of the Atlantic, female masculinity played a far less prominent role as a signifier of the modern woman in popular culture. Laura Behling has argued that in the United States, the literary, media, and scientific responses to female militancy in the years of suffrage activism ridiculed and pathologized the masculine woman, rather than ascribing any value or glamour to her, and Lillian Faderman has noted that no large lesbian subculture was established in the United States after World War I, as it had been in Britain.[2] In a similar vein this research indicates that there does not appear to have been the same degree of playful, oppositional, outrageous, or avant-garde elements in motoring women's public expression, as there was in Britain. American activists were more disposed to emphasize the liberatory potential of commodity consumption in their public campaigns than their British counterparts were, making for a distinct history of American women motorist's love of cars.

Frank Donovan began his 1965 history of American motoring, *Wheels for a Nation*, with a story of women motorists at the turn of the twentieth century. "Mrs. Astor's Horseless . . ." was the title Donovan gave to his opening chapter.[3] The women motorists he invoked, however, were rather shadowy figures, wheeled out to make a symbolic point about the class exclusivity of early motoring but given little substantive part in the subsequent development of American automobile culture. He wrote of ladies from New York's first families, ultra-privileged society matrons and their frivolous daughters, dressed in outlandish automobile fashions, who developed a mania for motor fetes and automobile gymkhanas at their Newport summer resorts. Unlike the motoring activities of men of their own class, women's automobiles parades, with their cars elaborately decorated with flowers and glow lights, and novelty competitions, such as spearing dummy pigs while driving, were represented as the epitome of lavish consumption. In marked contrast, Donovan cast the enthusiasm of their husbands and brothers for the early Gliddon trials, the Paris to Peking race of 1907, or the New York to Paris run of 1908 in much more honorable terms. Donovan told about the motoring exploits of male members of America's first families as stories of masculine inventiveness and technical experimentation, but frivolous society ladies, according to his commentary, remained incapable of understanding the mechanical gravity of their newest toys.

That differing emphasis in the gendered meaning of motoring served to set the stage for a particularly American version of automobile history, in which changing class structures and social democratization were visible and expressed through mass automobile consumption. Society women's automobile parades and lavish picnics provide a patently unjust and even laughable point of origin for narratives of rapid democratic change in the United States. They underscore a proudly nationalistic story of automobiles as a measure of twentieth-century democratic egalitarianism. The stories tell us little, however, about the variety of ways in which American women from a wide variety of backgrounds took to automobiles as the cost of cars fell within their price range. It was not until almost thirty years after Donovan's dismissal of them that American women motorists were given sustained historical consideration, in Virginia Scharff's *Taking the Wheel: Women and the Coming of the Motor Age* (1991).

The automobile was a European invention, and it was some years before manufacturers and consumers on the other side of the Atlantic began to take it seriously. The special contribution of American manufacturers was their conviction that automobile consumption was for everyone.[4] By 1904 production in the United

States had exceeded that of France and three years later was greater than France, Britain, and Germany combined. While many British and European manufacturers were still producing craft-built vehicles for prosperous middle-class families into the 1920s, American producers, led by Henry Ford, had cast the car as a gauge of populist democracy during the first decade of the century. Cars, which were becoming a valued item of consumption even for poor farming families in the West and Midwest by the late 1910s, rapidly moved beyond the bounds of extreme class privilege to be an indicator of American prosperity and social progress. The American car was hailed as a reforming force, measured by its distribution across social classes, capable of resolving social tensions without political conflict. As historian of technology Carroll Pursell put it, the automobile reinforced "deeply rooted values of individuality, privatism, free choice, and control over one's life. It was the perfect example of the nation's habit of trying to replace politics with technology."[5]

At the same time as stories about how female members of America's wealthiest families incorporated automobiles into their social rituals were filling the society pages, young women of somewhat lesser privilege were quietly beginning to think of cars in more than sporting or recreational terms. "The first woman to . . ." is a recurring theme in early automobile talk, and time after time the motoring press registered surprise at the numbers of women who demonstrated an active interest in cars. In 1900 a New York motoring magazine noted the arrival of the mechanically skilled woman Miss Eva Mudge, an "expert and ardent" chauffeuse "who knows how to drive an automobile anywhere it will go, and is not baffled by a short-circuit or a faulty contact."[6] Mudge was able to drive both steam and gasoline machines and declared that driving an automobile was much less dangerous than riding a horse. Numerous references to motorists such as Eva Mudge can be found scattered through American women's magazines, motoring journals, and newspapers in those early years.

Mrs. Dan Gaines of Colorado Springs, for example, had run a successful tour business from the turn of the century, driving visitors over mountain roads in her horse-drawn carriage, up the famous Pike's Peak and through the Garden of the Gods on moonlit nights. In 1906 *Automobile* noted, "She handles horses like a man, drums up trade at the railway station, has a trust on the theatrical business, pulls politics to a certain degree, and is a power to be reckoned with."[7] A photograph of Mrs. Gaines in her new White Steamer, the automobile she had bought to replace her horse-drawn hack, accompanied the article, but the convenient characterization of her as "the original and only woman hack driver" seems most unlikely. The impulse to identify an originating moment for women's desire to use automobiles as sources of income on similar terms to men — "the first woman to" — was always going to be a futile exercise. Women in various places conceived of such plans at much the same time, and, even more important, for some women such as Mrs. Gaines it was a logical transition from previous forms of employment.

That repetitive desire to locate an originating moment of female automotive exceptionalism deprived women's engagements with automobiles of a sense of continuity. Indeed, while appearing to inaugurate such a history, it was based on the assumption that women's technological competence had no past. The desire to identify the first woman to undertake a supposedly male role was not confined to male commentators, however, and the motoring journalist Mrs. A. Sherman Hitchcock, a tireless advocate for female motorists, used her columns to draw attention to women who were taking up driving as a profession. She often wrote of so-called first women, though she was less inclined than male journalists to make jokes at their expense. Instead, Hitchcock portrayed such women as admirably modern — handsome socialites and plucky college graduates who had turned their automotive skills to good account when their family fortunes had been reversed. She announced in 1910 that Miss Eleanor Lorraine Beattie of New York City "is undoubtedly the first young woman in the United States to advertise for a position as a chauffeur." Beattie had placed an advertisement in a New York newspaper stating that she had owned and driven several kinds of automobiles but was forced "through untoward circumstances to seek situation as chauffeur."[8] She could supply the highest references and expected only a moderate salary.

In the same year Mrs. Hitchcock reported with great approval Miss Katherine Lockwood's theatrical announcement, "I can take a machine apart and shake the pieces in a bag, and then put it together again." She quoted the young St. Louis woman's emphatic confidence in her skill at some length. Lockwood declared that she was fully qualified for a career in the motor trade. She could sell a car where a man would fail, was more likely to be able to get her automobile home without recourse to a team of horses than the average male chauffeur, and was convinced that many women would prefer to ride with her than with a male driver. "I think it is a splendid occupation for women," she declared. "All I ask is fair play, a chance, and I feel confident I will get it for I know a motor car from tire to top and from carburetor to clutch."[9]

Other women who articulated similarly strong views on their competence were featured in Mrs. Hitchcock's columns during the prewar years. Miss Jeanette Everett, a taxicabist from Philadelphia, confessed to being rather lonely on the streets with only male chauffeurs and hoped that more women cabdrivers would join her before too long. Hitchcock reported that it seemed that her wish might not be far from being realized by 1912 in Chicago, where twenty women were preparing to pass city taxi license examinations so they could work with the largest taxi company in the city. Feminists in New York supported professional women drivers, and in 1913 Mrs. Olive Schultz set up business in front of the headquarters of the Woman's Political Union. A male journalist who was invited to try her cab was impressed by Mrs. Schultz's skill with the apocryphal hairpin—a tool that, apparently, looked at home in a woman's hand. The reporter noted that Schultz was able to fix her car no matter what had gone wrong and could "do more to improve a peevish carburetor with a hairpin than most mechanicians can with a hydraulic drill and an assortment of monkey wrenches."[10]

Only a year later, quite forgetting Mrs. Schultz, the *New York Times* was once again hailing New York's "first feminine chauffeur," Miss Wilma K. Russey. She was employed by Dalton's garage at 354 West Fiftieth Street, where she had been working as a mechanic for more than a year. With a theatrical flair drawn from her earlier employment as a circus performer and with a fine appreciation of the performative aspects of her accomplishments, she wore high tan boots, long black leather gloves, and a huge leopard skin hat and stole. By 1916 a Miss Nancy Deanne was placing advertisements in New York newspapers, seeking work as a chauffeur. When interviewed, she declared, "It is hard to sit in an office day after day, to know all the time that you are as well qualified for the work you love as any man, and yet be unable to obtain that employment."[11]

Many such stories flourished during the war years. The press frequently reported women's wartime activities in Britain as well as the paramilitary mobilizations of American women even before Woodrow Wilson committed United States troops to join the war in 1917. As in Britain, new openings for women in automobile work were created as the military recruited male repair and service personnel into motor transport units, leaving shortages in civilian life as well as in the military services. Some, such as the van Buren sisters, granddaughters of the former president, toured across the country on their motorcycles promoting the pro-war "Preparedness Movement," in advance of the United States entering the war. Others made their way to France, sometimes with their own vehicles, to join voluntary ambulance units. Wartime service enabled privileged American women to frame their automotive passions in terms that were less open to social censure and quite outside of the connotations of frivolous femininity that continued to adhere to early motoring women. Rather than images of wealthy motor girls whose interests lay in leisured consumption and social frippery, American women were able to represent their activities as emblematic of high-minded patriotism and humanitarian service.

Press reports of American women's mobilizations at home and in France, however, differed somewhat from those in Britain. Unlike the expressions of collective feminist activity that had characterized accounts of British women's motor work, in which women's businesses were presented as seeking to provide employment and training for other women, reports of women's similar aspirations in the United States were generally framed in more individualistic terms. The press published articles about women's desires for automobile training and reported on their enthusiastic attendance at automobile schools, but in these stories men remained the expert teachers and women the pupils, albeit particularly adept and talented ones.[12] While there may have been some female trainers in the United States at that time, there appear to have been no schools set up "by women, for women," as there had been in England.

Similarly, American women's work in auto garages, both before and after the war, was described quite differently from its British counterpart. It was mainly within economic narratives of individuality, free enterprise, and the commercial exploitation of women's special qualities that women's involvement with automobile sales, service, and repair were most often appraised. Rather than feminist "ladies motor garages," as they emerged in Britain, with their avowed political, social, and economic dimensions, American garage women were celebrated for their business acumen and their admirable assumption of authority over the working-class men in their employ. Given that American motorists opted to drive themselves without the help of chauffeurs and that auto repair work quickly became formalized as a low-status, masculine trade, there was little space for privileged women to imagine an honorable or exciting role for themselves within it, as there was in Britain. The few accounts of women's automobile work in the United States during those first decades of the century emphasize the capacity of women to capitalize on their special womanly strengths in order to run exemplary businesses, rather than engage in mechanical work of their own. Garage women were noted for their ability to manage male employees as well as—and sometimes even better than—men, but, where the press attributed a more personally transformative or broader feminist motivation to a garage enterprise, it was only occasionally considered to be the kind of change that might bring women together or offer benefits beyond the individual entrepreneur.

Mrs. E. M. Self, the manager of a Delco garage in St. Louis in 1915, was featured in the motoring press as having achieved success in a masculine world by wielding a feminine broom through a failing business. Her success was not attributed to any mechanical skill. Rather, she was said to have rescued the garage by literally cleaning up the premises, eliminating inefficiency, and forcing "swearing, smoking and tobacco-spitting chauffeurs" to mind their manners, turning them into "alert gentlemen."[13] "The Female of the Species is More Efficient Than the Male: Woman Takes Charge of Man-Managed Garage and Puts Balance on the Right Side of the Ledger" was the headline that recorded Mrs. Self's achievement.

There are few press reports on women's garages in the United States in those years. They generally assumed the lowly status of mechanical work, and therefore journalists were not inclined to romanticize women's aspirations in that field. *Sunset Magazine* published a story about a "lady garage man," Mrs. L. W. Caswell, who grew up on a farm in Indiana and by 1913 was the joint owner and manager of a repair business with her husband in Saugus, California. The *Ford Owner and Dealer* featured a former manicurist, the owner of a Ford garage in 1923 on Manhattan Island at Park and 126th Street named Mrs. Rose Klein, who "as far as she knows is the only woman in the world who runs an automobile repair shop." *Automobile Topics* published a feature about Miss Margaret Neil and Miss Ruth H. Stockwell, who operated the Willys-Knight dealership in the small Iowa town of Marshalltown in the mid-1920s.[14]

Reports presented female garage proprietors as singular women — individual instances of women's mechanical aptitude, determination, and business acumen — who exemplified women's desires for increased employment options. Mrs. Caswell, the female reporter enthused, could "lift off the heavy hood in her strong brown arms as easily as if she were lifting a babe from its cradle." Rose Klein "had to struggle before she got to know every bolt and nut in a Ford, before she trained her ear to be delicately sensitive to its symphony of sounds and taught her hands to alleviate all its aches and pains." And Margaret Neil, who was "not afraid of grease or dirt," "is an adept in every mechanical feature of the automobile. She knows what it is to completely tear down and build up various types of engines."

While the reports expressed admiration for the women's determination and competence, they rarely asked how women might access mechanical training or raised questions about how women might enter the automotive field. Wealthy suffragist Rosalie Jones, interviewed in 1915 while she was undertaking mechanical training in a New York garage, provided something of an exception. She declared herself to be a pioneer, "breaking into" the automobile industry in ways that could provide openings for other women in the future. Society girls, she believed, should be trained for the business world, just as their brothers were. She was enjoying the training program, Jones told the reporter, and was considering entering the business of car sales, but, unlike upper-class suffragists in Britain, she expressly distanced herself from professional auto repair. "I would loathe the work were I to do it all the time," she was reported as having said. "I have no such intention. I merely want to know how it is done."[15] Like Hilda Ward and other women motorists of the previous decade, Rosalie Jones wanted to be sure that she would not be left dependent on the dubious skill of a male chauffeur or mechanic. Mechanical work as an everyday activity or as a way to earn her living held little glamour for her.

In the distinctive class and race context of American motoring, and without the robust discursive field developed through the women's right-to-work campaigns in wartime Britain, stories about women's garage work in the United States remained narratives of individual exceptionalism, which positioned women as successful small business managers and bosses over men. They were not inclined to place women's actions within a larger social project of collective feminist action, nor did they suggest the joys of transgressive gender blurring or female masculinity, as it emerged in Britain and Australia. Garage women's activities were not located within more radical considerations of women's desires that questioned the boundaries of femininity or redrafted notions of female bodily powers or within women's collective aspirations to instigate change in the gender order by embracing the social possibilities of a new technology. Instead, when American women's aspirations for social change were expressed through automobiles, it appeared in a very different form, one that utilized the links that had developed throughout the Progressive Era between commodity consumption and the democratization of social life.

It was stories of American women's pleasurable consumption of automobiles, rather than their aspirations as professional motorists, that dominated press accounts. In particular reports of women's enthusiasm for adventurous, transcontinental travel generated the greatest public debate. Manufacturers sponsored and promoted women's early transcontinental journeys in the years surrounding World War I, seeking to maximize publicity for their products and fostering new visions of personal identity that rested on automobile consumption. Their promotional material was distributed on a national scale, and it carefully located women's journeys within the two dominant discursive fields that surrounded modern American femininity in those years. First, it referred to feminist political aspirations, which were then being expressed through spectacular public parades, theatrical stunts, and other flamboyant incursions into masculine public space; and, second, it drew upon the exploits of Hollywood's intrepid serial queen heroines, with their deeds of bravery and derring-do in planes, trains, and automobiles.

The second decade of the twentieth century saw important changes in women's suffrage campaigning in the United States, and women's transcontinental journeys of the 1910s took place in that atmosphere of renewed right-to-vote activism. Four Rocky Mountain States had passed women suffrage amendments by the end of the nineteenth century, and after a doldrums of fifteen unsuccessful years, in which no states had adopted women's suffrage, feminist campaigns were finally beginning to make some gains. Success came only in the West, however, with five western states passing suffrage amendments in rapid succession after 1910. The victories galvanized the national movement, and suffrage organizations began casting about for new ways to conceptualize their campaigns, press their demands, and capture the attention of a mass audience.

Membership in suffrage groups increased during that second decade, and recruits across a broader social spectrum—not only socially prominent women but also working-class women and a new cohort of college-educated women—became active in a more heterogeneous and radical movement. They brought a fresh style to their campaigns, less constrained by notions of respectability, and they rejected the reliance on discrete, behind-the-scenes influence that had dampened women's activism in the nineteenth century. These changes fostered new tactics that exploited emerging forms of corporate advertising and mass consumption, as feminist activists learned to maximize their impact by drawing upon elements of modern consumer culture. Automobile campaigning, movies with a suffrage message, elaborate pageants that were a curious mixture of classical imagery and high-tech theatrical effects, outdoor rallies and parades, electric signage, and (toward the end of the decade) picketing and mass arrests were all employed with the media in mind. They brought new life to forms of politics that American men had largely stopped employing and helped to place woman's suffrage at the center of the national debate.

At the same time as feminists were taking to the streets with renewed energy, the Hollywood studio system was turning out tremendously popular stories of female daring and bravado in long-lived series targeted to a female audience such as *The Perils of Pauline* and *The Hazards of Helen*. With their array of brave and intrepid heroines, these films were built around the courage and athletic feats of stars such as Helen Holmes, Australia's "diving Venus" Annette Kellerman, Anita King, and Pearl White. Quite different from earlier stage melodramas, these silent cinema serials were characterized by realistic, on-location shooting; behind-the-scenes profiles of the stars that emphasized the real risks they took in their movies without recourse to stunt doubles; and action-packed plots that were frequently built around technologies of mobility—in trains, planes, hot-air balloons, and automobiles. Movie industry publicity blurred the lines between the stars' public persona and private lives, making the on-screen stunts appear all the more plausible. Helen Holmes, for example, was photographed for a feature article repairing her movie stunt car with her infant daughter sitting beside her in an empty tire.

When automobile manufacturers engineered press stories about women driving across the continent, singly or with other women, they were thoroughly immersed in the visual and narrative conventions of the serial queen genre. Women's transcontinental journeys were reported as tales of female bravery, athleticism, technical accomplishment, and "nervy" vitality that resonated with the heroines of the silent screen. The Maxwell Car Company began the succession of women's transcontinental crossings in 1909, when it enlisted the well-known reliability trial driver and president of the Women's Motoring Club of New York, Alice Huyler Ramsey, to drive from New York to San Francisco. Ramsey traveled with three companions, taking almost two months to complete the journey. They were billed as the first all-female transcontinental motorists, only six years after the first male crossing. The next year the Willys-Overland Company supported Blanche Stuart Scott, soon to become one of the country's earliest female aviators, and her journalist friend Amy Phillips to drive from New York to San Francisco. Their Overland car, dubbed "The Lady Overland," was painted white and outfitted especially for the tour with tools, spare parts, and wardrobe trunks. In the third and most publicized of all the trips, Hollywood silent screen star Anita King drove in the opposite direction, from the Panama Pacific Exposition in San Francisco in September 1915 to New York. Not only did King have the publicity department of her automobile manufacturer, the Kissel Motor Car Company of Hartford, Wisconsin, organizing and promoting her trip, but she also had the backing of the Paramount Pictures publicity machine. The two combined to bill her as "the Paramount Girl who drove Koast to Koast in a KisselKar." In 1916, in one of the last of the sponsored transcontinental tours before public interest waned, Amanda Preuss, a law office stenographer from Sacramento, drove an Oldsmobile roadster at a cracking pace from Oakland to New York to set a women's continental speed record of eleven and a half days. The Olds Motor Works published her account of the journey in a promotional brochure, *A Girl—A Record and an Oldsmobile* (1916).[16]

Manufacturers carefully and elaborately staged each of these trips for a national audience. The Maxwell Car Company presented Alice Ramsey with their latest thirty-horsepower touring car in New York, fitted it with a luggage rack and extra large gas tank, and provided fuel, mechanical support, and local guides through its network of agents and dealers along what was to become the Lincoln Highway. To orchestrate the publicity, the *Boston Herald*'s automobile editor was hired to travel by rail ahead of Ramsey's party. He telegraphed press releases and dispatches of their progress to automobile journals and newspapers throughout the country. The national press published numerous photographs of their journey—the party departing New York and arriving in San Francisco and leaving and entering small towns, Alice repairing flat tires on the prairie and negotiating rough dirt tracks in the western states, and occasionally their mishaps and breakdowns.

"A Girl, a Record and an Oldsmobile." Amanda Preuss, transcontinental motorist, drove from Oakland to New York in eleven and a half days in 1916. Courtesy of Sacramento Archives and Museum Collection Center, Eleanor McClatchy Collection, 1982/005/5782.

The Willys-Overland Company similarly supported Blanche Stuart Scott's journey as she zigzagged west for over three months, appearing at promotional events organized by Overland agents in many towns along the way. Scott telegraphed her location to dealers across the nation, and they updated her progress in their display windows, exhibiting her telegrams and moving a cardboard cutout of her car along a special strip map. Amy Phillips, Scott's companion, produced a booklet of the trip, which was distributed by the Willys-Overland Company as an advertising brochure, *5000 Miles Overland*.[17] As in Alice Ramsey's journey, a press agent was assigned to travel with them, and he carefully staged a great many publicity photographs, showing the slogans painted on the side of the car: "Overland in an Overland" and "The Car, the Girl and the Wide, Wide World between New York and San Francisco." Newspapers along the length of the route produced so many stories of their adventures on the road that Scott declared their journey resembled "one continuous parade," with carloads of photographers and reporters escorting them in style from town to town.

Anita King's 1915 crossing best exemplifies the close connections between real-life transcontinental trips and the Hollywood adventure serials. Her progress across the country was updated daily on a display board at the Panama-Pacific Exposition in San Francisco, and, as King traveled east, she delivered lectures about her adventures on the road, "Hazards of Helen" style, in more than one hundred Paramount theaters in the towns she passed through. Her travels were reported in newspapers across the country as thrilling action plot, often wildly embellished. She was threatened by a tramp in the lonely Sierra Nevada, was forced to shoot a coyote that was stalking her in the "Great American Desert," became lost for two nights and was rescued by prospectors, had to dig herself out of mudholes near Salt Lake City, received a proposal of marriage in Wyoming, and counseled starstruck girls wanting to run away to Hollywood about the dangers of the big city. When she returned to Hollywood, Anita King was given the starring role in Lasky-Paramount's "great automobile photoplay," *The Race* (1916), based in part on her transcontinental adventures and blending fact and fiction. A dramatic publicity still from the movie showed her apparently stunned and barely conscious in her smashed car after jumping it across a river, attesting to the physicality and daring of King's on-screen heroine.

Car companies sponsored women's transcontinental journeys to promote their products and services. The trips served to prove their car's reliability in extreme conditions; to highlight their national network of dealerships, no matter where their automobiles were produced; to elicit free editorial copy to turn their brand names into household names; and to show that their automobiles were, as advertisements for Alice Ramsay's Maxwell declared, "the car for a lady to drive—simply perfect and perfectly simple."[18] The women knew the deal and played their part. "I'm averaging 250 miles a day and the Olds car is the best on earth," Amanda Preuss told an Iowa reporter. At a tour of the KisselKar factory in Hartford, Wisconsin, Anita King declared that the most enjoyable part of her trip was the "daily companionship with my motor": "I grew more and more pals with my car."[19]

A central feature of the reports of women's transcontinental trips was their pleasure and pride in their capacity for independence and fortitude. "You should have seen us get the machine out of an irrigation ditch in Wyoming," Alice Ramsay, who claimed she was "born mechanical," told the *Sacramento Union*. "We just took out the block and tackle, hooked it to a stump at the top of the ditch and although it was hard work, we got the machine out all right."[20] That kind of effort would pass without saying when it came to male motorists. Men's stamina and resourcefulness were assumed and could be confirmed in a glance. But for women determining how to live that competence as they traveled through rain, dust, mud, and hot sun in an open car, without straying too far from the norms of femininity, was not an easy task. How they appeared, and not simply how they used the powers of the new technology, was crucial to the success of their enterprise. Alice Ramsey, traveling in 1909 with older female relatives, recalled that "the subject of clothes was a *very* important item" and needed a great deal of planning. The *Toledo Daily Blade* reported that Blanche Scott had been "besieged with inquiries from women all along the line as to the contents of their wardrobe boxes." Ramsey's and Scott's parties wore full garments still determined by Victorian restrictions—long skirts under heavy dusters or rubber ponchos and large hats with veils.

Only a few years later Anita King and Amanda Preuss, both working girls of much lesser privilege, could wear more masculine outfits of riding breeches, gaiters, khaki Norfolk coats, walking boots, and leather caps with goggles, though they also carried "dainty evening gowns" to wear in town. In common with press reports about the stars of the adventure serials, a great deal of effort was devoted to reassuring readers of the women's continuing femininity, in spite of their impressive feats. Reporters expecting to be confronted by intimidating Amazons, emphasized with relief how "dainty," "tiny," "pleasing," and "womanly" the overland automobilists remained. Journalists always took care to stress that, in spite of work that "would have taxed the strength and endurance of any man, they have appeared apparently guiltless of fatigue and gowned in correct evening dress."[21] In these accounts the women's strength resided not so much in their muscles but in their thoroughly modern nervy vitality, which allowed them to continue driving long after they were physically exhausted.

As the publicity surrounding the transcontinental motorists emphasized, automobility expressed individual women's aspirations for adventure and change. But their trips were also a forum for more collective, utopian dreams and suggested technological answers to the circumscribed mobility and vulnerability in public spaces that characterized even privileged women's lives. Alice Ramsey, expressing a determination to take care of routine repairs during her drive west, declared her intention "to demonstrate that women are independent of mere men and that they can and dare brave the wilds of various parts of the country, overcome many difficulties and take care of themselves."[22] Blanche Scott similarly expressed a desire to prove that "two girls can go anywhere without the protection of men and that a woman is capable of handling a gasoline car, even to fixing punctures, looking after the engine and putting on tires."[23] Even more explicitly, the motoring trade journal, *Motor Field*, used the language and imagery of suffrage campaigning to represent Blanche Scott's ambition. She was "willing to be a sort of Joan of Arc of motoring to lead her sisters into a campaign to demonstrate that women are not merely an ornament at the wheel of a car—that she is not to be confined to the simple child's task of driving a slow electric vehicle."[24]

In these terms independent women motorists came to be linked with women's suffrage during the years of strengthened feminist campaigning. Contestants in the reliability run in early 1909 organized by the Women's Automobile Clubs of New York and Philadelphia, including Alice Ramsey, were dubbed "suffragettes" in the motoring press, even though it was simply a recreational event. The very sight of a convoy of female drivers initiated a chain of associations that linked women's automotive and political aspirations.[25] Blanche Stuart Scott, quoted by Mrs. A. Sherman Hitchcock two years after her transcontinental crossing, drew similar links between women's aspirations to fly airplanes and their efforts to obtain the vote, noting that "women are in the forefront of nearly everything, whether it is vote-getting or volplaning [flying]."[26]

At the same time as automobile manufacturers drew upon images of modern resourceful femininity expressed in the adventure serial genre as well as through the language and practices of suffrage activism, the American suffrage movement increasingly found imaginative resources for its campaigns through images of a "consumer's republic" promoted by automobile manufacturers. These connections sometimes resulted in open collaborations between suffragists and car manufacturers. In 1914, for example, the Maxwell Motor Company enlisted the help of suffragists to advertise their new policy of hiring women sales and demonstration staff on equal pay with men.[27] Managed by the prominent suffragist Crystal Eastman Benedict, the program was launched in New York with speeches by other famous female voting rights activists, including Inez Milholland Boissevain, while a Barnard graduate, Jean Earl Moehle, dressed in a workman's leather apron and denim coat, stood on a platform in the Fifth Avenue showroom window for the afternoon, pulling down and reassembling a Maxwell engine.

While automobile manufacturers promoted automobile sales to women by associating themselves with images of female emancipation and occasionally with the organized suffrage movement, suffrage activists were incorporating automobile manufacturers' imagery of independence, mobility, and freedom into their political campaigns. That influence was most clearly expressed in the tradition of suffrage touring that came into prominence throughout the second decade of the century. There was nothing new, of course, about suffragists embarking on political pilgrimages. As early as 1877, Margaret Campbell traveled through Colorado for three months in a horse and buggy loaded with suffrage tracts, and a tradition of caravanning had been important in British suffrage campaigns. Numbers of American women—including Alice Paul, who was to become the leader of the Congressional Union (CU), the militant "ginger" organization that perfected automobile touring—took part in such tours during their visits to Britain.

Of all the American suffrage tourists the most flamboyant and enthusiastic was the New York heiress Rosalie Jones. During 1912 she joined the New York and Ohio campaigns for state suffrage amendments, traveling in a specially converted yellow wagon pulled by her horse "Suffragette." She soon developed more elaborate forms of political street theater, and that winter Jones led twenty-five suffragists on a march from New York City to Albany to present a parchment petition to the new governor. Atrocious weather and a jocular relationship with the press brought the petitioners widespread national coverage.[28] Front-page reports of the action evoked Washington crossing the Delaware and Pilgrim foremothers, associations that the marchers encouraged by carrying birch staves and wearing plain, hooded cloaks over their rather fine clothes. The next year Jones led a similar march for 230 miles from New York City to Washington, D.C., to join the suffrage street parade planned for President Woodrow Wilson's inauguration, organized by the Congressional Committee of the National American Woman Suffrage Association (NAWSA) and led by Alice Paul. This time the marchers devised even more elaborate rituals, including a service as they crossed the state line from Delaware, in which the "pilgrims" knelt and blessed Maryland soil in the name of suffrage.

"General" Rosalie Jones and her suffrage "army," as they were dubbed—articulate, well-bred, indomitable, disheveled, and covered in mud yet able to stop overnight in some of the best hotels—generated a great deal of free publicity. But it was publicity over which suffrage organizations had little control. The marchers' actions rested on backward-looking sources of political inspiration, a symbolic renouncement of twentieth-century modernity that worked against presenting the suffrage movement as a modernizing and progressive political presence, and their activities were liable to attract as much ridicule as admiration. Alice Paul, according to some newspaper reports, had greatly upset the marchers by sneering at their hiking garments and declaring their bedraggled condition a disgrace to the cause.

Shortly afterward, the suffrage movement renounced retrogressive references to pilgrim foremothers or the War of Independence and embraced automobile campaigning for its efficiency and as a way of generating images of suffragists as technologically sophisticated women of the future at the wheel of their own destinies. Political cartoonists were quick to draw upon the links that were being forged between women's political and automobilic freedoms. One showed confident young suffragists behind the wheel of an automobile able to burst through the impediments of elderly, masculine, machine politics. Rosalie Jones herself moved away from the anachronistic forms of public ritual she employed at first and in later marches exchanged her horse-drawn support vehicle for an automobile driven by Olive Schultz, the "suffrage taxicabist of New York," whose driving skills were greatly admired by the press accompanying the pilgrim hikers. By 1915 Rosalie Jones had become the proud owner of a yellow touring car, enrolled in a mechanics' course at a New York auto repair shop, and was considering taking up work as an automobile saleswoman.

Although the major American suffrage organizations all came to use automobile campaigning in the mid-1910s, they adopted different approaches. In the West, where rates of car ownership exceeded the national average, activists rapidly devised exuberant forms of automobile campaigning, and organizers attributed the success of their campaigns to such attention-grabbing tactics. The main umbrella group, the College Equal Suffrage League of Northern California, published an influential manual outlining precisely how it managed its campaign to get the suffrage amendment passed. Far from being "dignified and womanly," as conservatives suggested, the group's members argued that suffrage campaigners should grasp any legitimate means to attract attention and reach an audience. They found automobile speaking to be the best way to reach voters. It saved time, energy, and money, and, "in the great farming valleys of California, automobiles are ordinary possessions."[29] They quoted one voter who confirmed their arguments: "Madam, as soon as you stood up in that automobile I said to my companion, 'Woman Suffrage'."

In the eastern cities suffrage organizations were more ambivalent about the use of automobiles. There cars were as likely to flag class privilege and to highlight social divisions as to offer images of a more democratic future. After much debate on their use, moves were made to ban automobiles from most of the annual suffrage parades held in New York City in the early 1910s in favor of drilled, disciplined marching by ranks of women wearing standardized clothing. These attempts were not entirely successful either, and there were many jokes about wealthy suffragists marching in expensive tailored suits and the recommended thirty-nine-cent hats. Yet eastern organizations soon acknowledged and welcomed the value of automobiles as mobile speaking platforms. Cars offered some protection from

Rosalie Jones, New York heiress and suffrage activist, learning to maintain her yellow touring car, 1915. Courtesy of Library of Congress Picture Collection.

"No Use, Old Man, She's Coming Through!" The unstoppable modern woman and the campaign for the Ohio suffrage amendment, 1912. *Cleveland Leader*, 15 August 1912. © 1912 The Cleveland Leader. All rights reserved. Reprinted with permission.

hostile crowds and provided a practical way of leafleting numerous polling booths on election days.

Midwestern suffragists were among the first to embrace the art of the massed automobile parade, and in the national capital Alice Paul, of the Congressional Union for Woman Suffrage, had come to embrace the symbolic and practical value of automobiles. She organized a national petition campaign in the summer of 1913, which culminated in eighty automobiles parading up Pennsylvania Avenue to the Capitol. By 1916 the less-militant NAWSA had also come to appreciate the promotional opportunities of transcontinental touring when two of its New York activists drove a yellow Saxon roadster around the United States to publicize the association's upcoming national conference in Chicago. Mrs. Alice Snitjer Burke and Miss Nell Richardson spent six months driving more than ten thousand miles through almost every state of the Union, using the car as a mobile billboard and speaking platform.[30] The Saxon car company used photographs of the two women's journey in its advertisements, endorsing women's suffrage at the same time as it promoted its own product.

For both automobile manufacturers and suffragist organizations, consumption suggested a fresh site in which women were defined, and able to identify themselves, as modern civic actors. In the "democracy of the road" they could become new kinds of subjects, expressed through the pleasurable consumption of mass-produced goods. As *Motor* journalist Margaret R. Burlingame mused in July 1919, just before the suffrage amendment was passed, the motorcar had been an important factor in the long fight for suffrage victory. She named prominent suffrage leaders who were enthusiastic automobilists and included a photograph of NAWSA leader Dr. Anna Howard Shaw, a woman who "not only uses a car, but drives it and repairs it," crank-starting her car. Feminists had relied on cars in their campaigns to secure the vote, Burlingame declared, and now the "score between the automobile and the woman bids fair to be settled" in the expanded sales the industry could expect from millions of newly emancipated women.[31] Suffrage organizations would be key to educating and even managing the mass of women into enthusiastic consumers, and the "automobile salesman goes on his way rejoicing, knowing that while there are women there will be automobiles."

At the same time as suffrage campaigners were able to claim enhanced, machinic powers that boosted women's claims to new forms of female citizenship through their use of automobiles, the material objects of cars themselves were changing. They were taking on new meanings in response, it was frequently said, to the particular desires of female consumers. The idea that car manufacturers were obliged to design cars with women in mind—"the feminine influence on cars," as it was usually called—was a notion that had become commonplace among observers of the automobile industry from the beginning of the century. Each year, as new models were launched, commentators declared that women's interest in motoring had forced manufacturers to produce cars that were ever more attractive, comfortable, safe, clean, and simple to operate. In 1912, the year that electric self-starters were first introduced into production cars in significant numbers, the White Car Company of Cleveland produced a clear statement of that feminization of the technology in their advertisement for their enclosed "Town Car," which they had "built particularly for women." It "offers the touring radius and flexible speed of the gasoline roadster, combined with the comfort, ease and safety of the electric brougham," read an ad in a Chicago arts journal in 1912.[32] The White car boasted features that soon came to be considered perfectly standard in gasoline cars, such as an electric starter, lights that could be operated without leaving the driver's seat, and a curbside door.

By the 1920s it seemed that most of the advances in automobile design were attributed to the special demands of female consumers. Technological developments, large and small, which tempered the rugged masculinity of early car design—from enclosed cars, fabric instead of leather upholstery, self-starters, automatic gear shifts, choice in paint finishes, glove boxes, four-wheel braking, rear-view mirrors, windscreen wipers, improved jacking systems, and tires that were easy to change—were said to have been a response to women's desires for automobility. Journalist Margaret Burlingame had declared in 1913 that women were the driving force in automobile sales and that refinements had been added "solely at the demand of the ladies." In pithy shorthand Madelaine G. Ritza summarized women's productive role in automobile consumption when she declared in 1921 that such changes were essential to the mass diffusion of automobiles as a practical, everyday technology: "Man made the automobile; women tamed it. Man utilized it; woman has socialized it. Man has commercialized the automobile, but woman has merchandised it."[33]

Ritza was restating assertions about the ways in which male engineers had been forced to adopt fresh approaches to automobile design in response to women's "natural" physical inferiority and "innate" aesthetic sensibilities. Such observations rested on chains of association that posited comfort and ease as particularly feminine desires, physical weakness and technological timidity as inherently female characteristics, and pleasure in refined and stylish consumption as a specifically womanly predisposition. From the point of view of men such gendered understandings bolstered a comforting investment in traditional ideals of masculinity placed in opposition to the commodification of mass culture, by suggesting that women were easily duped by fripperies of styling and fashion in automobiles, in contrast to men's rational adherence to solid engineering values.

Implicit in those definitions of women as particularly irrational and needy consumers, however, was a simultaneous acknowledgment of the ways in which the nuts and bolts of cars came to embody women's desires for substantive social and political transformations. Stories about the feminine influence on cars undercut the disjunction between masculine production and feminine consumption by defining women as creative consumers whose desires for personal automobility had the effect of transforming utilitarian, masculine machinery into everyday social objects. Women welcomed the capacity of cars to provide pleasure and physical challenges within a personalized "zone of safety," or mobile "living room," that enabled individuals and small parties of travelers to move out of their homes with a gratifying degree of insulation from the dangers and inconveniences of public spaces. As women invested cars with their ambitions for vaporizing the limitations imposed on their mobility, they simultaneously enhanced their authority as technological actors and consolidated their status as citizens with a stake in the life of the nation.

Manufacturers' efforts to sponsor and publicize the activities of women motorists did much more than simply sell cars. They were an important element in developing a new discursive field in which both automobiles and femininity acquired fresh meanings through the connections being made between them and in which ideals of female citizenship were expressed through the language of commodity consumption.

CHAPTER 5

Campaigns on Wheels
American Automobiles and a Suffrage of Consumption

In November 1915, only three weeks after yet another defeat of a state referendum giving women the right to vote, a dispirited New York suffrage movement was offered a different course of action. A West Coast activist named Sara Bard Field, who (unlike her eastern hostesses) already had full voting rights, was advocating an alternative to the state-by-state campaigning that had so far dominated women's suffrage politics in the United States. She stood in the sumptuous Sherry's ballroom at the Plaza Hotel in a travel-worn brown suit and, as the *New York Tribune* put it, appealed to her Fifth Avenue hostesses. "We want to help you, we voting women of the west," she pleaded, "will you let us? Will you take our hand in the fight for the federal amendment?"[1] Field had traveled all the way from San Francisco by automobile to hold out that hand to the women of the East. She told them she had shivered in the Rocky Mountain passes, starved in the Nevada deserts, and spent a long night in a Kansas mudhole to reach out to them. But the real heroines of the hour, the reporter declared, were two women sitting at the back of the stage, their rough fur coats and warm caps in stark contrast to the fashionable audience. They were Miss Ingeborg Kinstedt and Miss Maria Kindberg, the owners, driver and machinist of the "sturdy little automobile" in which Sara Field and a 500,000–name petition had traveled. They were the ones who had made the suffrage meeting possible.

That high-profile automobile campaign of 1915 reveals a great deal about the ways in which feminists in the United States used automobiles to promote their political aims. Sara Bard Field, Ingeborg Kinstedt, and Maria Kindberg's actions were very different in style and emphasis from British suffrage campaigns, as their journey was mounted at a time when most of the British women's movement had put aside campaigning for the vote in favor of participating in the war effort. The war had altered the forms in which British women's political aspirations were expressed, though it did not end their demands for suffrage. The militant arm of the British suffrage movement had turned into a fiercely patriotic, militaristic organization and had lost much of its membership, but the bulk of British activists put their efforts behind supporting women's paid and voluntary participation in war work, expecting that doing so would demonstrate beyond doubt that British women deserved the vote. The suffrage automobile campaign of 1916 was distinctively American in its emphasis. Through it women's mobility, their engagement with new technologies, their pivotal role as consumers, and their visibility in public places was crucial. Activists' deliberate conjunction of two key elements of the modern experience—consumer culture and independent women in public spaces—worked to amplify both, inflecting femininity and automobiles with new associations in ways that were specific to the American context.

Automobiles helped suffragists to move out of their parlors and hired public halls and into the streets. But, as the many florid newspaper accounts of the "suffrage envoys" and their three-months tour from the West Coast to the East suggest, cars were more than just practical tools for facilitating political campaigns. It was not only speakers and a "monster petition" that their "sturdy little automobile" carried from San Francisco to New York but a great deal of symbolic freight as well. As the momentum of publicity grew for the suffrage party driving across the country, the women and their car together became a public spectacle in which new kinds of female citizenship were graphically enacted. Suffrage campaigners' political aspirations meshed perfectly with manufacturers' commercial interests, with manufacturer's advertising narratives of coming social equality to be delivered through mass consumption.

The winning of universal suffrage in the United States was not one of steady forward movement, as progressivist histories so conveniently assume. Instead, the political gains momentarily won by black men with the passing of the Fifteenth Amendment in 1870 were rapidly and systematically lost to them.[2] Jim Crow provisions, which created state constitutional loopholes such as literacy tests and property restrictions, as well as tactics of intimidation and sheer terror, guaranteed formal and informal impediments to African-American men's enfranchisement. Knowing that many black men were being denied the voting rights supposedly secured for them by the Fifteenth Amendment, black suffrage activists—men and women both—focused on the political goal of unimpeded universal adult suffrage, rather than on women's suffrage, which was the limited goal of the white

Automobile campaigns for suffrage, a democracy of commodity consumption. *Literary Digest*, 16 October 1920. By permission of the National Library of Australia.

movement. Numbers of black women, however, did campaign for women's suffrage in the years leading up to the passage of the Nineteenth Amendment in 1920. They worked in separate black women's organizations, such as the Alpha Club of Chicago, and at times sought inclusion in white organizations and campaigns, in spite of being rebuffed. As Rosalyn Terborg-Penn has pointed out, African-American women recognized that, on the occasions when white women did invite them to take part in joint actions, it was on strictly circumscribed grounds and partly in an attempt to attract the black male vote in favor of female suffrage. White women's suffrage organizations, though they differed on many issues, were in agreement in their refusal to engage with issues of racial equality. It was not until the civil rights movements of the 1960s and beyond—two generations later— that an approximation of genuine adult suffrage was achieved in the United States.

While the major white suffrage associations concurred on questions of race, there were major differences between them on other matters. Disagreements between the largest body, the National American Woman Suffrage Association (NAWSA), and its more militant offshoot and rival, the Congressional Union (CU), were reflected in a geographic separation in the spheres of influence between the two organizations.[3] During the mid-teen years of intensive state campaigning in the eastern states, NAWSA made it clear that the CU was unwelcome in the East. The CU responded by building a power base in the enfranchised states of the West and Midwest. Its plan was to bypass painstaking state campaigns and to organize women who had already gained the vote into a bloc that could apply political pressure on legislators in Washington for a federal amendment. It was a controversial and divisive plan that broke with the women's suffrage tradition of nonpartisanship. Democratic women were urged to vote against all Democratic candidates in order to persuade Woodrow Wilson's administration to pass a federal amendment. The tactic was easily characterized as a Republican plot, even though many of the CU activists, including Sara Bard Field, were avowed Democrats. There is no evidence that such a bloc was actually mobilized, and Wilson's Democratic government was reelected in 1916, but the tactic helped to create a sense of urgency around women's suffrage, and it brought suffrage activists to a clearer understanding of party politics and strategy making.

In the course of their campaigning in the western states, the CU devised a platform and rhetoric that would distinguish it from the older style of the NAWSA. One of the most powerful rhetorical devices the organization hit upon relied on the frontier mythology so closely associated with the American West. But the CU's version reversed that mythology's familiar trajectory. Instead of portraying men on horses riding toward the sunset to claim new territories—that fraternal, masculinist image of nation building—it was to be a narrative of newly enfranchised women in automobiles driving back toward the dawn, bringing their youthful and wholly nativist energy to the corrupted and ethnically heterogeneous political life of the old cities in the East. Theirs was a vision of female solidarity, in which politically empowered western women were returning to the "dark lands" to help their weak eastern sisters. In her farewell speech in San Francisco, Frances Joliffe, the other "envoy" of enfranchised women selected to travel with Sara Field, employed the metaphor of slavery to represent the situation of eastern women, which she added to that reverse frontier imagery. They were traveling to states where unenfranchised women were "enslaved in the factories and mills," working backbreaking hours and unable to register their protest. They were heading east, armed with the fighting strength of four million western women voters, to help set those women free.[4]

Automobiles, with their promises of autonomy, freedom, flexibility, individualism, and self-determination, were ideal vehicles for the expression of that modern feminist twist on manifest destiny. For a society in which liberal ideals rested on notions of free people in command of their own actions, the trip provided the media with stories that tapped into familiar national themes. A Baltimore newspaper declared, "The prairies, the Rockies infested with 'Injuns' and coyotes, impenetrable forests, valleys, streams and rivers have had no terrors for four women in an automobile who are traveling hitherward from San Francisco." The *Washington Herald* adopted a similar rhetoric. Schoolboys of the present generation read thrilling tales of the Daniel Boone "He Killed a Bar" type, but, the paper proclaimed, soon schoolchildren would be learning how a carload of women had "chugged, speeded and skidded their way eastward over mountains and plains, through the sands and marshes, with a six-cylinder for their charger, and an automobile toolkit for their sort of modern crusader's spears."[5]

The imagery was suggested by the enormous publicity surrounding the Panama-Pacific International Exposition held in San Francisco in the summer of 1915, from which CU's transcontinental automobile trip was launched. The exposition was presented as a celebration of the vitality of the West and of the power of twentieth-century American technology to meld the nation into a new unity. It marked the completion of the Panama Canal, a massive engineering project that provided a new link between the East and the West, and announced the resurgence of San Francisco after the devastation of the 1906 earthquake.

In the years before the exposition and even more so after the outbreak of war made a summer tour of Europe impossible, it was predicted that a flood of eastern visitors would come to the West for the first time. Cruise ships offered special excursion fares via the canal itself to the fairgrounds in San Diego and San Francisco, but most visitors traveled cross-country on railway packages especially tailored to the exposition visitor and first-time tourist. The road lobby had also been active in those years, and the Lincoln Highway Association, which had been formed in 1913, rushed to transform the ill-defined trails of the northern transcontinental route into a passable road in time for the exposition, publishing optimistic accounts of the fine conditions that drivers would encounter.[6]

In fact, the Lincoln Highway was a highway in name only, in no way complete when the exposition opened. Only a few isolated "seeding" miles of road had been paved, and, in the exceptionally wet spring of 1915, sections west of the Mississippi were more accurately described as an endless series of mudholes. Emily Post, who was sent by her editor at *Collier's* magazine to report on the exposition and the conditions she found on the road, traveled via the Lincoln Highway in a luxury touring car. She had expected something like the *routes nationale* of France but instead found "a meandering dirt road that becomes half a foot of mud after a day or two of rain." An *Official Guide to the Lincoln Highway* was hastily published for exposition tourists, but it contained numerous errors, and the red, white, and blue route markers, which were painted on rocks, trees, fence posts, barns, and telegraph poles by local civic associations eager to encourage business, were sometimes washed away by the heavy spring rains. The reality was that western sections of the "highway" had been assembled from an informal network of local roads where the regional traffic was greater than the transcontinental traffic. It confused strangers because the busiest-looking road was not always the throughway and could just as easily end in a mineshaft or at a ranch.

Discouraged by the older suffrage organizations from entering the eastern campaigns, the CU had set up a lively information booth in the Palace of Education at the San Francisco Exposition. CU leaders had traveled to San Francisco to build a base from which they could reenter the East with their call for a federal campaign, once the expected suffrage amendment defeats in New York State, New Jersey, Pennsylvania, and Massachusetts had occurred. They enlisted the help of San Francisco activists, and the booth quickly became a dynamic focus for feminists who visited the exposition. A program of public lectures and cultural events was scheduled, offering a national and international focus on the suffrage movement. Displays—from dolls dressed in the national costumes of suffrage nations to information about the voting behavior and suffrage sentiments of state and federal politicians—were exhibited at the booth and expanded throughout the course of the exposition. As witness to that energy, they compiled a giant petition in the form of a roll of paper that draped down a wall and onto a table. Voting women were invited to sign the petition demanding a federal amendment enfranchising women, the "Great Demand" of the Congressional Union, also called the Susan B. Anthony Amendment. By September the CU claimed the petition contained a half-million signatures and was a half-mile long. It was taken to be a material sign of women voters' growing awareness of their new political power and became something of a sacred object, which the suffrage envoys were to display and enlarge on the journey east. Their plan was to present it to President Woodrow Wilson in Washington, D.C.

The CU's work at the exposition culminated in a national conference held over three days in September 1915 and ended with a flamboyant pageant, designed by the New York producer, Hazel MacKaye, who had been brought over especially for the event. Ten thousand women packed into the exhibition grounds to see them off with suffrage tableaux and specially composed anthems sung by a massed choir. The CU's timing was impeccable, and the transcontinental party left San Francisco in a surge of high emotion that was widely reported in the press. Earlier, throughout July and August, various newspapers had noted plans for a petition pilgrimage from the exposition to the national capital that would include scores, even hundreds, of women on horseback and in trains, ships, and automobiles. But in the end only one car, a modest Overland Six, drove out of the exposition gates to begin the long journey to the East. It carried the two envoys of the women voters of the West, Sara Bard Field and Frances Joliffe, and two suffrage activists from Providence, Rhode Island, Ingeborg Kinstedt and Maria Kindberg, who acted, respectively, as "mechanician" and driver.

Instead of beginning their journey, however, the four women went home to bed, and it was not until some nine days later that the party quietly slipped out of San Francisco on their way to their first meeting in Sacramento. Mabel Vernon, who had orchestrated the successful Nevada suffrage campaign in 1914, took on the job of advance agent. With CU leader Alice Paul, Mabel Vernon went ahead by train to prepare rallies for the envoys in Reno and on the steps of the capitol in Salt Lake City. Frances Joliffe, a wealthy San Francisco socialite active in the labor movement, pulled out of the trip at the last minute, citing ill health, though privately Sara Field wrote that Joliffe was bored by the two Swedish women and became appalled as she came to appreciate the hardships of the trip ahead. Joliffe, however, promised to travel by train to meet them in the East at the end of their journey. Many newspaper stories continued to include Frances Joliffe as a member of their party, even though only three women—one envoy and the two automobilists from

Campaigns on wheels. Suffrage envoys sent by the female voters of the West to lobby for a federal amendment. Courtesy of the National Woman's Party Collection, Sewall-Belmont House and Museum, Washington, D.C.

Providence—were left to go on. That false start to the trip, which was picked up by some of the press, was an early sign of the CU's appreciation of the importance of carefully orchestrating their symbols and rhetoric for maximum publicity.

The need for a carefully managed publicity campaign was a subtlety that Maria Kindberg and Ingeborg Kinstedt, both naturalized American women of Swedish extraction and both over sixty years of age, were not entirely at ease with because it misrepresented who they were. In the interests of a broader narrative coherence "the Swedes," as they were generally referred to in private correspondence, were forced to acquiesce to a style of politics that relied on their silence, even though the success of the trip was dependent on their money, their automobile, and their technical skills. It became apparent as early as the farewell ceremonies in San Francisco that the driver and mechanician were to be cast as heroines but mute ones. They would be on display and spoken of in admiring terms but rarely permitted to speak. For, even though the two believed they had plenty to say and despite the fact that radical women in the United States admired the progressive social policies and feminist critiques that had been developed in Sweden, the pair were greatly at odds with the imagery that the CU was attempting to project. Kindberg and Kinstedt lived in the East and so were not voting women; their strong accents betrayed their nonnative origins; their age made it difficult to represent the campaign as an expression of fresh, youthful energy; their eccentric, old-fashioned clothing placed them far from the desired image of female modernity; their ambiguous sexuality and their radical, syndicalist politics and bluntly expressed opinions on the shortcomings of American men and the hypocrisies of American social arrangements made them loose cannons on a journey that was made under the scrutiny of the press. In short, the two Providence women were far from the image of West Coast youthful glamour that the CU encouraged. The CU's campaign publicity rested on images of independent and adventurous all-American "girls" who blew in from the West, "as fresh as debutantes," offering their wholesome energy to the enervated political world of the East. And, even though Sara Field herself scarcely fit that image, the plausibility of the story relied on keeping the two older women firmly in the background.

Without Frances Joliffe, Sara Field became the focus of the trip. In her early thirties, a poet and a socialist with an extensive background in western suffrage campaigns, Field was an inspired speaker whose assurance and passion grew throughout the months of the campaign. Of her arrival in New York and the reception at Sherry's ballroom, she wrote to her lover in Portland of the hundred cars that met them in Central Park and escorted them to the ballroom. Joliffe had joined them, as promised, but her speech was "wretched." Field, on the other hand, built on her experiences along the way. "I held that rather tired and rather blasé audience spellbound, as I blazed before them this newer, more militant aspect of suffrage. I had them laughing with me and later crying, and finally made the appeal for a solid East of women to meet a solid West."[7]

Ten days later, when the petition was finally handed to President Wilson at the White House with an even bigger parade, her grandest moment out of many, Field was able to present herself with all the authority of someone who had participated in, even invented, new versions of a nation-building ritual. By surviving the trials of distance and extreme landscape and carrying evidence of her contact with people right across the nation, she claimed she had dissolved some of the differences between men and women and had placed herself on an equal footing with the greatest office in the country. But, in spite of such triumphs and the prominence it brought her, Sara Field's commitment to that long trip also caused her a great deal of personal grief and hardship.

At that time she was in poor health, emotionally fragile, and plagued by economic insecurity. Field needed to earn money to help support her adolescent son and daughter living in Berkeley. They had been placed in the custody of their father following a bitterly contested divorce, which she had obtained through a long residency in Nevada. She divorced her husband in order to live with her lover, Charles Erskine Scott Wood, a veteran of the "Indian wars" and now a radical Portland barrister, freethinker, and poet. Wood, however, was prevaricating about leaving his wife and his other complicated affairs in Oregon to join her in San Francisco. The emotional intensity of her scandalous relationship with Wood added a compelling energy to her oratory, and the enthusiastic responses she elicited along the journey lifted her out of the illness and despair she experienced while waiting for Wood to make his promised move. Her health, it seemed, was never better than when she was on that trip. As for the opportunity to earn some cash, Field had been asked to send dispatches of the trip to the *Bulletin* in San Francisco and the *Portland Oregonian*, but the difficulties of typing in a jolting car and the relentless speaking schedule left her too exhausted to write. Halfway through the trip she begged Mabel Vernon and Alice Paul to have Frances Joliffe brought over to take her place, but she had become too valuable to be let go.

The sheer length of the journey, four thousand miles in ten weeks with constant moving over long stretches of rough, unsealed road, was extremely demanding. In Nevada, where no rain had fallen since June, the roads were dusty and corrugated with potholes and washouts. They "ploughed rather than rode" through the thick dust and were "jolted into hash," Sara wrote home. The party sometimes became lost on the poorly marked route and was forced to drive through freezing desert nights to arrive in time for the meetings Vernon had arranged for them. Further east, sections of the road were quagmires, which some motorists claimed had been deliberately maintained by local farmers as moneymaking ventures. A front-page story in the *Emporia Gazette* of Kansas, headed "Three Lone, Lorn Weemin" and treasured by Field late into her life, told a story of exhausted, muddy militants arriving at midnight and sneaking into the Mit-Way Hotel: "They got on famously until they struck Nickerson and got stuck in the mud. They had no man along, and it was a woman's job to get them out. Miss Kindberg is the driver and Miss Kinstedt is an engineer, and having no man along to do the swearing, it was sheer strength and moral courage that got them out."[8]

The women did not simply follow the Lincoln Highway but zigzagged through as many towns as possible. By the time they were in Des Moines, Alice Paul had secured the coup of having the envoys address a joint session of Congress, to be followed by a meeting at the White House with Woodrow Wilson, so the focus of their travel became to arrive in Washington at 11:30 A.M. on 5 December 1915. With the promise of such a high-profile finale, the CU accelerated its publicity campaigns in the eastern states, organizing even more meetings and pulling up wherever the traveling party could draw a crowd.

Mabel Vernon, a consummate and experienced campaigner, acted as tour manager, advance agent, and liaison with headquarters. She traveled ahead by train, lining up press interviews, arranging for politicians to meet the party, and trying to drum up a parade of local suffragists to welcome the campaign car's arrival in each town. Sometimes Vernon had only hours to organize a reception before the overland party arrived, but the coverage she generated was enormous. The four Salt Lake City papers, for example, each published a story every day for four days preceding the women's arrival, and they all gave front-page coverage to their rally on the capitol steps. Even in small towns where the travelers stayed for only a few hours, local presses carried prominent stories about the trip and outlined the CU's tactics to secure a federal amendment.

In towns large and small the routine was the same. Mabel Vernon would arrange for local suffragists sympathetic to a federal amendment to meet the travelers on the outskirts of town. Banners and bunting were hung on the Overland Six, and whoever was available would parade to the city hall, the courthouse steps, the capitol, or the governor's mansion, where the mayor, the governor, senators, and representatives were pressured at a public meeting to declare their positions on a federal amendment and add their signatures to the petition. Afterward, prominent local suffragists hosted private receptions at which the CU's tactics were explained, a new branch formed, and funds solicited toward campaign expenses. Sara Field and Mabel Vernon were the star speakers at these meetings, and Ingeborg Kinstedt and Maria Kindberg were given the job of selling subscriptions to the *Suffragist*, the CU newspaper. In towns such as Cleveland, Ohio, where suffragists were loyal to the NAWSA, the travelers were given the cold shoulder and had to fend for themselves. When that happened, Vernon simply hired a band, called on the local Overland dealer to loan a car or two, and made a show anyway. "You would have laughed," she wrote to Alice Paul of their Omaha visit, "to see the splendiferous way we carried it off all by ourselves."⁹

The CU's plans were made on the run, as the short-term planning and contrary announcements in the press reveal. Money was always a serious problem, and poor communications between the party on the road and Alice Paul, the ultimate decision maker at headquarters in Washington, made the logistics of a transcontinental campaign a major challenge. Long-distance telephone calls were still a rarity and well beyond their budget, and the daily letters and telegrams between Mabel Vernon and Alice Paul show plans being devised and altered as they traveled. Vernon was continually begging for names of possible sympathizers in the towns ahead. Headquarters in Washington, in turn, wanted news clippings and reports of their meetings in towns they had just left, to publicize the car's progress. They claimed that the CU could place any number of stories and photographs in the eastern papers, but the party had usually left town before the papers appeared, and the dispatches Sara Field occasionally managed to write while they were on the road usually arrived in Washington too late for publication. Other problems, large and small, needed continual attention. The petition—ostensibly the whole focus of the journey—went missing; their car broke down outside Topeka, and Mabel Vernon spoke for two hours to hold the crowd before she had to give up; Sara Field's love life threatened to become a public scandal in Detroit; suffrage rivalries in the towns they passed through needed to be carefully negotiated; Vernon's personal bank account, which floated the expedition, was frequently overdrawn until Alice Paul could wire funds; and at various times all three of her charges threatened to pull out of the campaign and head home. "No one will know how thankful I will be when I see them safely into Washington," Vernon wrote from Des Moines.[10]

At that time Field could not drive. She declared she was not "mechanically minded" and was totally dependent on the two Swedish women to see her through. "The one who is driving the car has delivered some 2,000 babies in her life-time and can probably deliver me safely into Salt Lake City," she reassured her lover as they set out from Fallon on a six hundred–mile desert crossing.[11] The two Rhode Island women, though unfamiliar with the territory, exuded confidence in their ability to get to Washington. In her early letters home Field described them as a "fine pair," "stalwart," "most trustworthy and agreeable" companions, and "the finest ever." From Utah she wrote to Wood that he would find them fascinating. They were really great women in the deepest sense of the word—free thinkers, devoted advocates of free motherhood, lovers of any form of liberty, and rebels against all the evils of our so-called civilization. Kinstedt, the mechanician, was full of humor, pithy comments, and wise observations. The driver, Kindberg, was not as talkative or expressive but was to Field more lovable and motherly.[12]

Their major failing, according to Field's account at the time, was Kinstedt's volubility. She was a "terrible talker" who talked "all the time," but Field judged it to be not as bad as it might be because she was also very interesting. Not surprisingly, that pleasure in her companion's forcefully expressed opinions wore off, though there is no hint at the major falling-out she would recall many years later. References to "the Swedes" simply drop out of her letters home after they pass through Kansas City. In an interview for an oral history project fifty years later, however, Field remembered Ingeborg Kinstedt in much more sinister terms. The mechanician had only just been released from a long stay in a mental asylum, she recalled, and had been belligerent throughout the whole trip, at one point even threatening to kill her once the journey was over. Field suggested that Kinstedt felt they were being treated as menials and was resentful that so little of the limelight went their way. Whatever the precise exchanges between the women in those three months of being thrown so closely together, Sara Field's memory of the events emphasized Alice Paul's single-minded persuasiveness when it came to convincing activists to fall in with her vision, whatever the personal cost. In her determination to get a job done, Field wanted to point out, Paul would accept any offer without bothering to investigate it and showed little concern for the dangers to which she exposed the campaigners. The story from the point of view of her two companions was, of course, quite different.

Even though the Congressional Union was inclined to keep them in the background, much of the press warmed to the driver and mechanician. They were variously described as "sturdy," "capable," "full of grit," "game ladies over sixty," "picturesque," and "dauntless, fearless women who shrink from praise for their courage."[13] The Providence press noted approvingly that not only was Ingeborg Kinstedt familiar with machinery, but she knew enough about carpentry to have built a bungalow. Part of the fascination was their appearance. The press noted their heavy boots, rough fur coats, strange headgear, and the old-fashioned dress worn by Kinstedt that was reminiscent of women's turn-of-the-century cycling outfits. The *Kansas City Post* called it "a remarkable khaki costume which may be buttoned one way to form a divided skirt and another way to look like a very full feminine garment."[14]

In newspaper reports that give a sense of the women's own words, their lively character is palpable. When they hired a man on the advice of locals to drive them across the Utah desert, he got them hopelessly lost but still demanded payment, holding them at gunpoint. Unscrupulous garage proprietors charged them excessive prices for fuel and stole their tools. Farmers demanded five dollars a time to pull them out of mudholes, though the women relished telling how they once recouped their money by subsequently charging the farmer five dollars to pull his mule out of the same mudhole with their car. "Everywhere the people soaked us," Ingeborg Kinstedt was quoted in the *New York Tribune*. "When they see suffragists coming they double the price. Ah yes, they are very polite, the American gentlemen, but they soak the suffragists."[15] As Maria Kindberg told the *Nebraska State Journal*, "We were told to provide red pepper to protect ourselves against dangerous men, but the only protection we have needed has been for our pocket books."[16] The press, of course, loved stories that suggested such wonderful headlines: "Messengers Bearing Tidings of Suffrage across Country: Only Needed Mere Man Once and He Proved Much of a Failure"; "Envoys for Suffrage Arrive after Exciting Adventure: Getting Lost in Desert and Following Trail in Blinding Snowstorm among Experiences of Four Pleaders for Equal Rights"; "Suffrage Envoys Stuck in Mudhole, but Came on Time"; and "Women Lose Way No Meeting Here: Two Punctures, Losing Their Way, and Rotten Roads the Men Build Blamed by Women for Delay."[17]

The CU provided very little information about the two Swedish women. How they came to have emigrated to the United States, their understanding of their role in the campaign, even their names are in doubt, as newspaper and private correspondence used various versions and sometimes anglicized the spelling. The *Providence Sunday Journal* identified them as active suffragists in their hometown, being the treasurer and president of the Women's Political Equality League, a group more militant than the moderate Rhode Island Equal Suffrage Association. Ingeborg Kinstedt was a member of Mrs. Kent's "Committee of One Hundred," charged with raising funds to finance the CU's headquarters in Washington, and was the suffrage correspondent for the *Searchlight*, the journal of the Rhode Island Society for the Suppression of Vice. A photograph in the Library of Congress collection shows Maria Kindberg as a suffrage "newsie," selling copies of the *Woman Voter* on the streets of New York.[18] Their technical proficiency, however, was their major value to the campaign. As Sara Field emphasized at the time and in oral histories fifty years later, the two women's competence was a crucial component in the success of the campaign, not just for practical reasons but because it provided a graphic and newsworthy demonstration of a new kind of modern, female citizen.

The suffragists' capacity to travel across country without the aid of men established that men and women were not "made of two separate kinds of clay," as Sara Field put it. "We didn't need any men to help us drive across the country, through all kinds of hardships and into situations that would have tried any man," she frequently declared to the press, implying that women should not be considered different kinds of political citizens either.[19] But the terms in which her admiration for her companions' automotive skills were reported simultaneously suggested Field's ultimate lack of interest in, and even regard for, such mechanical matters. A Des Moines reporter quoted Field's somewhat offhanded admiration for their skills when the party had a breakdown in the wilds of Nevada. When they discovered that someone had stolen their tools, "Miss Kinstedt did not say anything, but she pulled the car apart, took out something and made a jack of it, raised the car, repaired the damage and replaced the something and the car went on." Field recalled campaign tours in Oregon when the male driver was unable to repair a breakdown. "It goes to show what a woman can do," she concluded.[20]

Given such resourcefulness, it must have been all the more galling to the driver and mechanician—and perhaps even the source of Ingeborg Kinstedt's remembered outburst of anger toward Sara Field—that, when newsreel footage was shot in Chicago, it was Field and not the Swedish women who the cameras focused on. She wrote to Wood with amusement, telling him that if he went to the movies he might see her in "funny trumped-up" Hearst and Pathé newsreels jacking up the car, "repairing" a puncture, and waving to crowds from the steps of the art institute.[21] That the two Swedish women would be so ignored, not only when it came to speechmaking but also in the very area in which only they had the expertise, must have been hard for them to bear. "We are a symbolic crew making a symbolic journey," Sara Field told a crowd in Lincoln, Nebraska, thereby eclipsing the realities of the actual journey.[22]

Ingeborg Kinstedt hinted at her mixed feelings of pride and annoyance at being considered a curiosity, at the same time as she could feel let down by those whose job it was to help. When they were in remote places, she would make her own repairs, but in the cities she did not, as it attracted too much of a crowd. One night in Providence she was forced to lie on her back in the snow under the automobile, because the garage man was eating his Thanksgiving dinner.[23]

The Overland Company promoted the women's achievement in driving their car across the continent, though only after the party had arrived safely in the East. The head office in Toledo fed dealers copy to place in their local papers, and numerous articles were published under headlines such as "Suffragettes Select an Overland Car" or "Suffragettes in Overland: Long Distance Tour in Aid of Their Cause."[24] The accompanying text detailed the car's qualities as well as outlining at length the political aims of the campaign. Suffragists clipped and pasted the advertising copy into their campaign scrapbooks, carefully preserving them alongside conventional news items and editorials. Those connections between automobile manufacturers and suffrage organizations blurred distinctions between commercial and political imperatives, linking consumer choice and political life in ways that were coming to sound natural. In manufacturers' promotional material, suffragists' choice of automotive products blended commercial and political languages, so that aspiring political citizens, suffragettes, were simultaneously invoked as consumers whose demands for political power were identified and even fused with their freedom to choose between a variety of commercial goods. Manufacturers' promotional material contributed to an emerging ethic of democratic consumption in the United States that has been called a consumers' republic, in which women aspiring for political rights were characterized as consumer-citizens.

As the story of this suffrage campaign suggests, the links between commodity consumption and female suffrage were not simply imposed by commercial manufacturers upon an unwilling feminist movement, undercutting and trivializing its serious political aspirations, as some analysts have been inclined to argue. Rather, the CU's transcontinental campaign of 1915 indicates that suffragists made deliberate and strategic choices to employ, exploit, and extend the power of the associations being forged between narratives of political citizenship and consumer culture. The way the trip was represented by the CU was designed both to enlarge and utilize the political dimensions of commodity consumption and women's roles within it. The organization carefully drew upon the connections that were being established between women as consumers and fantasies of equality delivered through new technological products. That their trip was successfully completed without the help of men gave a material fullness to suffragists' demands for political equality and provided them with a fresh register through which to demonstrate women's new status. "If you have any doubts as to woman's ability, possibly as a voter," wrote a journalist in Kansas, "talk to Miss M. A. Kindberg of Providence Rhode Island who drove the little car all alone through the wind and rain and sun. Miss Kindberg won't say much, because she believes in deeds instead of words."[25] Their battered Overland car was given pride of place in countless parades and was said to be the object of close scrutiny by men and boys.[26] The sight of the car, with its worn banners, luggage roped onto the running boards, and the western mud and dust carefully preserved to heighten its effect, allowed them to stage a highly stylized performance of female independence, transcontinental solidarity, and sexual equality. It was a nonverbal vocabulary of entitlement and social change, perfectly intelligible and easily read by people in the streets.

Yet, as CU campaigners worked to incorporate automobiles into their political rhetoric and practical campaign repertoires, they both broadened their perspectives about how they might create new forms of citizenship and limited their vision about which people might be entitled to assume that citizenship. While their attempts to exploit the identifications between political entitlement and consumer goods attracted a great deal of publicity and helped develop new forms of campaigning, they simultaneously acquiesced to the exclusions and inequalities on which that political vision rested. The CU's campaign tactics implicitly imposed limitations on a movement avowedly concerned with injustice and inequality. Not everyone stood in the same relation to automobiles, after all, and, despite optimistic fantasies, the car by itself could not refashion social relations. Rather, it was an elusive and complex combination of automobile, particular kinds of people, and a specific historical circumstance that gave the campaign its layers of significance. Even Maria Kindberg and Ingeborg Kinstedt—the two women who actually owned the car, possessed the skills to drive it across the continent, and stridently declared their belongingness to that new polity—were only tenuously and reluctantly accorded the status of automobilized–female citizens to come. Those who were placed even farther outside the new middle-class patterns of commodity consumption could be entirely erased from the language of political entitlement that was being created.

As suggested by the CU's resident cartoonist, Nina Allender, in her piece published in the *Suffragist* of September 1920 with the caption "Any Good Suffragist the Morning After," that circumscribed vision of citizenship made it easy to believe that the campaign for universal adult suffrage had come to an end with the ratification of the Nineteenth Amendment. But, rather than representing *any* good suffragist, Allender's image of a campaign successfully completed showed only one very limited version of feminist identity, the "Allender Girl," as she had become known.[27] It showed the archetypal new consumer, a young, middle-class, white woman, contentedly sleeping in, newspapers with headlines announcing the passing of the women's suffrage amendment discarded around her bed. For those women whose ethnicity, race, and class easily placed them as the "right type" to claim membership of the political body and where their sex constituted the only impediment to political entitlement, a victory had indeed been won. But for others the struggle to gain a political voice, not to mention the opportunity to lie in bed as they pleased, had only just entered a new phase. Black women found that, because state legal loopholes had already been put in place to block black men's access to ballot boxes, the voting rights conferred on them by the Nineteenth Amendment were even more quickly removed than they had been for the men before them. When they turned to white women for help to counter the rapid moves to disenfranchise them, African-American women were rebuffed by a movement determined to separate issues of race from what they argued were "purely" women's rights issues. White feminists' disinterest in addressing the injustices of race illustrates how the modernity they desired, even though it was couched in the language of inclusiveness, was grounded in a racial ideology. Suffrage organizations that could so readily harness their visions of women's emancipation to consumer culture—democracy represented by, and measured through, the widening dissemination of material goods—were poorly placed to conceive of success in other ways.

CHAPTER 6

"The Woman Who Does"

A Melbourne Women's Motor Garage

British and American women embraced automobiles as a technology that would expand their lives, although their actions and how they were represented in the motoring and popular press were not the same in the two countries, each reflecting a distinct national context. Australian women also loved cars in the 1920s. It is important to note that Australia became an independent nation only at the beginning of the twentieth century, when the six separate British colonies federated into the Commonwealth of Australia, just as the first motorcars were being imported. Thus, Australians took to the automobile at a time when the colonial experience was still historically and emotionally close to the national consciousness. That colonial context is important to understanding the specifically Australian meanings that motoring acquired and how Australian women were able to embrace automobiles for both pleasure and profit in the early years. Their emphasis, consequently, was quite different from that of women in Britain and the United States.

In the United States the role of the railway in molding the disparate states into one nation has been well documented. From its logistical importance in the Civil War, its central place in a characteristically American industrial revolution based on agriculture and transportation, to its role in opening up the western frontier during the second half of the nineteenth century, the railroads have long been identified as a crucial vehicle of national integration. In Australia, however, that moment of national unification occurred just as automobiles were becoming a viable form of transportation. By the end of World War I, when the population of the United States had already exceeded one hundred million and Britain had over forty million people, Australia had a population of less than five and a half million, mainly living in a narrow strip along the southeastern coastline. With a landmass similar to that of the United States, transport was a major problem in Australia, and the nineteenth-century technology of coastal shipping remained the most important form of long-distance transportation well into the twentieth century. There were few waterways suitable for inland shipping, so Australians relied on walking, bicycles, and animal power (horses, donkeys, camels, and bullock teams) for decades after other industrialized societies had adopted mechanical forms of transportation. The first railway lines were built in the 1860s, but they were predominantly local lines. A transcontinental rail route from Sydney in the East to Perth in the West was not completed until 1917, and even then it was only a single track of different gauges that were not standardized until 1970. A north-south line between Darwin and Adelaide was not completed until 2003. Consequently, it was not the railroads but automobile technology that bypassed parochial state rivalries and provided Australians with a modern transportation system that could unify the continent into one nation.

Insecurity about Australia's remote frontiers permeated national life, so that transportation linking the city and the bush was of much more than utilitarian value but had major symbolic importance as well. Travel was difficult between settlements, and the huge tracts of land scarcely inhabited by Europeans, but patently home to Aboriginal people, appeared as a constant reproach to the incipient nation's claims to occupy the entire continent. Australia's status as a European outpost—geographically placed on the edge of densely populated Asia but unable to secure its borders or make productive use of its resources—imparted a sense of fragility to the new nation. Indeed, there was considerable debate about whether the northern parts of the country were suitable for white settlement at all. The preoccupation with the task of fully possessing the continent was expressed in a persistent fascination with the unpopulated areas—the "back of beyond" of European settlement, the "bush," the "outback," the "never-never," and the "top end" of the tropical North. Settler Australians' sense of identity, contrary to the realities of coastal urbanization, leaned heavily on images of the bush and on stories about places that most people would never visit.

During this period of national formation private cars and commercial trucking acquired an immense importance, not only in practical ways but also because they told much about what kind of nation Australia had been and was to become. Automobile technology could "break out fresh country" for settlement and tourism by placing the stamp of civilization and white settlement on the "trackless wastes" and give travelers from the cities a sense that they were taking part in a pioneering venture. Trucks and cars helped to call into being a modern Australian society, finally ready to become up-to-date, its citizens able to hold their heads high in the company of the Western world.

Although the country did not develop its own car industry until after World War II, Australians embraced automobile consumption with enthusiasm. From the earliest days rates of motorcar ownership in Australia were among the highest in the world, behind the United States but well above Britain. The first half of the 1920s was a particularly expansive period. Measured against the average wage, the initial purchase of a car in 1925 was less than at any time until 1965, though high maintenance costs in the interwar period meant that ownership was largely confined to prosperous middle-class families. Statistics are not available for women's car ownership or driving licenses in this period, but women were prominent in press stories about the ways in which automobiles were transforming life in Australia, especially in remote areas.

There was a firm and frequently asserted belief that Australian women displayed a particular type of modern femininity, quite different from British and American women, which reflected the historical circumstances of colonial life and even the characteristics of the landscape itself. Non-Indigenous Australian women had won the right to vote at the turn of the century, among the first in the world to do so, and observed the ongoing suffrage campaigns of British and American feminists with some sense of superiority. Indeed, white Australian women were often held up as models of advanced female citizenship. Practicality, physical courage, independence, familiarity with the bush, and ease within a masculine world of action were the qualities commonly attributed to them. That image of modern Australian femininity found expression in the bush heroine characters of fiction and narrative cinema, such as the "Billabong" children's series written by Mary Grant Bruce in the 1910s and 1920s; in silent era feature films such as *On Our Selection* (1920), *The Breaking of the Drought* (1920), and *A Girl of the Bush* (1921); as well as in Australian women's own assessments of their difference from women in the "old world." Stories of adventurous and resourceful young women gave expression to a colonial femininity that downplayed the importance of sexual difference and emphasized young women's delight in gender equality, technological expertise, and demanding physical activities— characteristics that were taken to indicate Australian women's readiness to step into the modern world. In a country where formal political equality already applied and where there was no significant manufacturing industry, images of Australian women motorists were not associated with suffrage. Nor were they simply used to promote the pleasures of mass consumption, though, like in Britain and the United States, a great deal of advertising copy employed images of glamorous women motorists. What was particular to the Australian context was the belief that women motorists' access to modern technologies enabled them to take part in the nation-building project, in ways that went beyond the usual female contribution of childbearing. Normally considered a masculine enterprise, the sense of urgency in constructing a new, white society in an old country promised women a degree of latitude and public recognition when they sought to move into new territory.

The aura of resourceful Australian femininity surrounded a prominent garage business, the Alice Anderson Motor Service, established in Melbourne, Victoria, after World War I. From 1919 until Alice Anderson's early death in 1926, the all-female staff of the garage were the favored "machinists" and "chauffeuses" of wealthy eastern suburbs households and taught countless Melbourne women how to drive and how to maintain their cars. "Most people have seen her neatly uniformed chauffeurs leap briskly from the driver's seat, open the door and salute smartly," declared one Melbourne newspaper.[1] The garage lost its high profile after Anderson died, though the business continued to be managed by women into the early 1940s. It closed when the staff—by then mostly women who had never known Anderson—left for military service in the next war.

Anderson's adult life, as her sisters recalled, was dogged by scandal. She was suspected of the full range of sexual transgressions—of living with men, of having abortions, and of being a lesbian. Her dramatic death at the age of twenty-nine of a gunshot wound to her head only days after she returned from a camping trip into central Australia with Jessie Webb, the first female history lecturer at Melbourne University, only served to create more gossip about her. While no hint of scandal emerged at her inquest and a verdict of accidental death was returned, persistent rumors of suicide brought about by financial worries or an unhappy love affair circulated for many years.[2] The coroner found, however, that a faulty pistol she had been cleaning caused her death in the workshop of her garage. Ironically, friends had urged her to carry the firearm for protection from "wild blacks" as the two women traveled through the Central Desert. News of her tragic death at such an early age attracted a great deal of attention. Melbourne newspapers placed it on their front pages and followed up with reports of her funeral and inquest. They published testimonies to her "career full of romance and rare achievement" and reported on the crowds at her funeral. Hundreds of wreaths were sent to Lancewood private hospital, where her body had been taken, and the press reported that her staff in the "dark brown trousers and leggings of their uniforms" carried her body to the graveside.[3]

Alice Anderson was born in 1893 into a professional Anglo-Irish household impoverished by the 1890s depression. Her family had been forced out of middle-class suburban Melbourne by the property bust to live in a rough bush hut on the outer fringe of the city. Her father, a civil engineer with Fabian inclinations, was frequently away doing contract work in remote shires. When Anderson's sisters recalled her early life, they told a story that was remarkably close to the sense of adventure that characterized colonial bush heroine plots. Alice had been allowed to "run wild," they said, dressed for practicality and economy in sailor's shirts, boy's boots, and bloomers. After their older brother Stewart died in 1913, her life came to resemble even more closely the adventure fiction she read so avidly. She took over the outdoor work of the household, hunting for rabbits, fishing, and tending the animals. As her sisters told it, she grew up strong, fiercely independent, and precociously competent. A favorite family story has the thirteen-year-old Alice saving the life of a timber cutter in a remote sawmill camp by stitching his slashed throat with a thread drawn from her horse's tail. Years later, with an awkwardness of expression that suggested the absence of a simple way to refer to a technologically competent woman, Anderson declared to *Woman's World*: "I always loved the outdoor life, and before I left home I was the handy man around the place. I suppose mechanics has always fascinated me."[4]

At this time middle-class women in Australia were increasingly entering professional training at the university level, but Alice Anderson's formal education was limited by her family's finances to the local half-time state primary school and a few terms in the Anglican girls' grammar school in the city. She did manage, however, to acquire a thorough grounding in the mechanical arts. Qualifications in automobile work were only beginning to be formalized in the 1920s, and that flexibility allowed her to receive mechanical training from sympathetic men, including her father. In 1913, in exchange for keeping the books at the community transport cooperative set up by her father, she learned to repair charabancs, the heavy open buses of the day, and to drive them across the unsealed and notoriously dangerous roads of the Dandenong Ranges surrounding their home. At eighteen Alice acquired her first car, an American seven-seat Hupmobile, as a birthday gift from her father. It was a large, open touring car with a fabric hood that he had bought for the transport cooperative, but the board of directors had refused to take on the debt, so he gave it to Alice with only the deposit paid. Alice was left

The Woman Who Does

Give a Girl a Spanner. Alice Anderson, motor mechanic, chauffeur, and proprietor of the Kew Garage, Melbourne. *Home*, 1 December 1920. By permission of the National Library of Australia.

to raise the 350 pounds needed to complete the sale—a major challenge when female clerical wages were only about 80 pounds per year.

Alice Anderson found an office job in the city and worked after-hours to pay for the car. She took touring parties on weekend picnics into the bush she knew so well; chauffeured young women to dances; conducted shopping tours for country visitors; drove women to hospital for their confinements; joined the convoys of Melbourne motorists who met troopships returning with wounded soldiers from the Great War; and continued her mechanical training at a city garage. Within four years Alice Anderson, then aged twenty-two, was able to leave her office job to open her own motor garage. She was quoted in a women's magazine using the same words as women in Britain and the United States to express her pleasure in professional automobile work, "I got the opportunity to vacate the office stool for the wheel—and I took it."[5]

Although she only worked from the backyard of the house in which she lived in a rented room, the advertisement Anderson placed in the *Directory of Victoria* of 1919 was bold and buoyant:

MISS ANDERSON'S MOTOR SERVICE
(KEW GARAGE.) Tel. HAWTHORN 2328
67 COTHAM ROAD, KEW.
Seven-seater HUPMOBILE and Five-seater DODGE Touring Cars for Hire.
Driving and Mechanism Taught.
PETROL, TYRES, and all Motor Accessories Stocked.
REPAIRS to all Classes of Cars.[6]

Her optimism paid off, and a year later, with the backing of some wealthy supporters, she had borrowed enough money to place a deposit on a nearby block of land. The local motoring press reported her scheme in sympathetic terms. "Miss Anderson, who has taken up motoring as a profession," noted the *Australian Motorist*, "has made an unqualified success of her venture. Her garage work in Kew, Victoria, has grown so rapidly that a three storeyed brick garage is to be built to her special requirements."[7] The magazine's account of her business plans called to mind the high-minded "New Woman" residential communities of the late nineteenth century. The first floor of the garage building was to house the automobiles that were in for repairs, the second would be a workshop, where only girls would be employed, and the third floor would be used for the sleeping and eating quarters for the staff.

That ambitious plan was never realized, but Alice Anderson did manage to build a single-story brick garage on a prominent corner in the main street of Kew. There her imagined live-in community of female workers was reduced to a small bedroom for herself, set into a corner of the garage. For the next seven years, until her death in 1926, newspaper and magazine articles reported the business in approving terms. "Possibly no woman in Melbourne was better known," wrote one daily newspaper. "She pioneered the way to motor garages for women and made a greater success of it than most men could."[8]

Anderson was familiar with women's motor garages in England during World War I. She read Australian press articles on British women's war work, corresponded with friends serving in Europe, and had access to British feminist magazines as well as to the motoring press and professional engineering journals. Australian women's moves to form military units on the model of those in Britain came to nothing, but British women's success in entering military service helped to foster a transnational climate for women's motoring ambitions. Always a staunch advocate of motoring as a suitable career for middle-class women, Alice Anderson claimed there was a growing demand for the trained "girl driver" in private service and expressed even greater optimism over the prospects for skilled female auto mechanics. A qualified woman mechanic, she declared, could earn five pounds per week and more, a good professional income for women at that time, though it is unlikely that any of her staff were ever paid that amount.

Anderson's vision of the Australian female professional motor driver and mechanic, through drawing some inspiration from women's military mobilization in Britain, was somewhat different in its expression. It was inflected not only by the distinctive role automobiles played in Australian national life and the flatter class structure of Australian society but also by the widespread belief in the superiority of colonial female resourcefulness over more hide-bound and socially conservative British manifestations. Australian women were particularly suited to the work, she told the press. Immigrants who had applied to work with her lacked the initiative and confidence of Australian girls—they were inflexible, familiar with only one make of car, unable to learn the streets of Melbourne, and had no knowledge of mechanical repair work.

To secure her garage workers' difference from, and even superiority to, male mechanics and chauffeurs, Anderson emphasized their social similarity to their customers. She kept a classical Greek textbook in the glove compartment of her rental car and studied it while waiting for clients. Even though she had not managed to complete high school, she expected extraordinarily high standards of professional skill from her female mechanics. Bright, educated, intelligent young women fresh out of college made for the best trainee mechanics, she declared. Any girl could be trained to be a good driver in a year, but, if she was to qualify as a thoroughly competent garage assistant, she needed to be able to repair cars and make spare parts if necessary, and that required an eight-year course, which included trigonometry, chemistry, and mechanics. "My ambition is to turn a trade into a profession for women, and it is well within the grasp of those who have initiative and grit," Anderson stated in her emphatic way.[9] These standards were much higher than that expected of male mechanics, as by the 1920s mass production and a growing spare parts industry were making workshop-based manufacture obsolete.

Like most garages of the period, the Alice Anderson Motor Service provided an eclectic range of services. It had a well-equipped mechanical workshop; ran a driving school, which included mechanical instruction on cut-away engines; offered a twenty-four-hour uniformed rental car service; and provided drive-in service and petrol sales. Vehicles were stored for clients who were abroad, and wealthy Kew residents used the staff as part-time chauffeurs, garaging their cars in her premises and calling drivers out by telephone as they were needed. The women drove stock and station agents on tours of inspection through rural areas, chauffeured wealthy clients to their holiday properties, and organized tourist excursions, in which Anderson's personal charm, bush knowledge, and cooking skills were a major draw. In 1926 she advertised a three-week tour to Sydney and the Blue Mountains and back to Melbourne for the huge cost of thirty-five pounds per passenger.

That broad range of activities constituted a serious bid for a slice of Melbourne's growing automotive service and repair industry. Alice Anderson's garage was registered with the Automobile Chamber of Commerce under several categories of membership, and she was among the first mechanics in Melbourne to apply for the newly established A-Grade Certificate, though she died before she could sit the examination. Anderson formulated grand plans for expanding her business, far beyond her financial means, into forms of tourism and travel that would not be realized in Australia until the post–World War II decades. She spoke of developing a fleet of motor caravans for hire to tourists and even of getting her pilot's license so that she could extend her business into aerial tourism. Alice Anderson wrote a monthly motoring column called "Her Wheel" for the magazine *Woman's World*, in which she did not engage with any of the contemporary press debates about the dubious skill of female motorists but provided matter-of-fact technical advice to a motorist gendered, without comment, as "she."

While her garage had much in common with other motor garages of its day, Anderson conceived of her business as much more than a foray into a technological domain wholly defined by men. Far from merely copying established business practices, Alice Anderson was a tinkerer, a creative participant in a growing and open field. She gained some fame as an innovator, particularly for what she called her "Get-Out and Get-Under" device, a foldaway creeper, or platform on wheels, that allowed the stranded motorist to propel herself under the car without having to lie on the ground. In a familiar Australian complaint, however, Anderson claimed her idea had been stolen and patented in the United States before she could take full advantage of it. Other inventions disappeared without trace. The *Australian Motorist* offered her patented "radi-waiter," a small tank fitted with taps that could be clamped on the rear of a radiator to keep beverages hot during a day tour, as evidence of her "inventive brain."[10]

Anderson attracted a great deal of free publicity for her garage with her innovative "once-over for thirty shillings," a service in which the customer's car underwent a complete overhaul by her team of mechanics in eight hours. Appreciating that women experienced difficulties in acquiring mechanical knowledge and confidence, she also developed services that were especially designed to attract and support women motorists. The *Australian Automobile Trade Journal* noted her novel idea, in which women car owners could work on their cars with the help of the garage mechanics, learning how to wash, service, and repair their cars in the process.[11] In a more ambitious plan women could attend the workshop as paying pupils for months at a time. They were supervised by the garage mechanics, in a kind of minor apprenticeship or finishing school in the arts of technological modernity. The scheme proved to be popular among girls leaving school, and prosperous parents, especially those from country regions, found it a suitable discipline for their automobile-struck daughters.

Lucy Garlick's father, a civil engineer in the far north of the state, sent his daughter to the Alice Anderson Motor Service for the summer of 1926–27, when she was just sixteen. Lucy learned to drive "the correct way" and was trained in basic repairs before she was allowed to drive him through the remote parts of the state for his engineering practice. Jean Robertson, born into a wealthy western district pastoral family, signed up for the motor maintenance course at the Alice Anderson Motor Service when she left school in the early 1920s. She became a regular contestant in amateur automobile sports events and drove a Riley 9 overland from Melbourne to join the Monte Carlo Rally of 1932. Marjorie Horne, a young Melbourne woman from a less wealthy background, was a student at the motor service for a full year in 1925, when she turned seventeen. She remembered the strict discipline and severe perfectionism of her training at the garage. She used her training to become a professional motorist, finding work as a private chauffeur and opening her own small mechanical workshop in a nearby suburb in the late 1920s.

Alice Anderson employed about eight women when her garage was at its height in the mid-1920s, and perhaps twenty women were trained as mechanics throughout the two decades of its existence. Some, such as Alice Anderson's younger sister Claire Fitzpatrick, who worked at the garage on weekends while she was an engineering student at Melbourne University, and Marie Martin, who began work as a junior trainee in 1924 at the age of sixteen, have left written records of their time at the garage. Marie Martin had also been born into a downwardly mobile professional family, part of Melbourne's middle-class bohemia, and, like Alice Anderson herself, Martin had none of the college qualifications that the proprietor had declared so necessary for the trainee girl mechanic. It was Marie Martin's first paid job, and she worked there until she was nineteen, when she married a mechanic employed at a rival garage across the road. In the early 1980s, in response to a research inquiry, Martin wrote in terms reminiscent of girls' serial fiction that she loved the work because it "appealed to her spirit of adventure," despite the long hours and low pay.[12] Like the paying pupils, Martin also recalled the strict professionalism of the training she received. Her first job was to dust and clean an opulent car with velvet upholstery and crystal vases, which was stored for a customer under a tent canopy in the back of the garage. By eighteen she had earned her driver's license, become a competent mechanic, and was chauffeuring well-to-do Kew matrons around Melbourne. She recalled driving clients to picnics, the races, and on shopping tours and even occasionally drove a regular customer the sixteen hundred miles to Sydney and back. Photographs show her, diminutive and slender, almost lost in a uniform of breeches, cap, leggings, and oversized woolen greatcoat, proudly holding open the door of a rental car for an imaginary customer.

In the postwar years conservative elements of the Melbourne press, as elsewhere, equated modern femininity with widespread social decadence, new female vices, and women's reluctance to become wives and mothers. That judgment was not universal, however; some women's magazines, motoring journals, and popular daily newspapers took a more progressive stance toward the ethic of experimentation and female independence that the garage represented, reporting on the Alice Anderson Motor Service in admiring terms. Although their optimism about her business demonstrating the demise of occupational barriers for women was premature, they recorded a fascination with the garage as an expression of modern femininity, never failing to mention the women's masculine style. They particularly focused on the clothes the women wore at their work, their hairstyles, and their breezy confidence. Alice Anderson's trousers always attracted attention, and she was characteristically described as a "boyish-looking figure in dungarees, generously ornamented with very real grease." She had an air of "youthful capability and the cropped hair she tosses out of her eyes suggest a sturdy self-reliant spirit."[13] A motoring magazine reported with approval the businesslike style she had adopted: "Skirts and hair have gone; she has donned male attire, and a woman's chief worry, her hair, has been cut closely, and she will pass now for a youth of 18 or 19 years of age."[14] Anderson's single-minded devotion to her business proved that she and her "khaki-clad chauffeuses and mechanics" could run an exemplary garage.

Such admiring reports about the garage and the women who worked in it drew links between that enterprise and earlier feminist campaigns. It was "out of true loyalty to her sex," as one put it, that Alice Anderson staffed her garage with female mechanics and pupils.[15] The very image of a residential community of professional, single women recalled the New Woman communities of the turn of the century—the settlement houses, nursing schools, and colleges that had been at the forefront of the battles for a previous generation of feminist activists. When a women's magazine described her garage building as "a roomy modern garage with sunlight streaming through lofty windows," it similarly suggested a modern version of older female communities, religious, medical, collegiate, or military.[16]

Yet the notion of collective feminist action is not the best way to understand Alice Anderson's garage. The ideals that underpinned her enterprise are better characterized as the kind of radical individualism often invoked by women who came to maturity in those interwar years. Like other women of her generation, Alice Anderson was inclined to emphasize the importance of individual accomplishments unrelated to gender difference and to act boldly, as if sexual equality had already arrived. If her garage was a feminist enterprise, it was an implied, noncollective kind of feminism and one that did not seek to claim that name. She presented her actions as an example of individual initiative in which she and other singular women creatively sidestepped the irrational restrictions of the time, rather than publicly campaign against them. In these terms her garage can best be understood not as an enterprise seeking to enhance the social status of women but one built on wanting to minimize or neutralize the importance of the category altogether. That repudiation of oppositional feminist action and the downplaying of sexual difference went together with expressions of femininity that no longer shied away from female sexuality—indeed, emphasized its importance in women's lives and even facilitated the emergence of a lesbian identity.

While girlish fantasies of mobility, adventure, and sartorial experimentation could be tolerated and even encouraged in fiction writing and cinema, it was an entirely different matter for women to build an ongoing public identity and a viable business around such identifications. The work provided the garage women with opportunities for improvisation and pleasure, but it also brought with it a degree of risk. The line between admirable gender experimentation and "aping men," as it was frequently termed in the years around World War I, was sensitive and uncertain, with the signs of demarcation in flux. For grown women to appear in public in breeches and leggings was a risky enterprise. It could be taken as evidence of commendable modernity and spiritedness, but it might just as easily slide over to signal the disreputable side of life. Marjorie Horne, one of Alice Anderson's young trainees in 1925, recalled the risk that inhered in their distinctive clothing. She wore boots and breeches for her work and remembered being viewed with hostility by some people in the streets: "I suppose they thought I was some mannish old thing . . . a he-woman," she told me. As Horne indicated, the problem for Anderson and the other women at the garage was just how their "tomboyish" desires for mobility, technological competence, independence, and pleasure might be translated into sustainable lives for grown women.

Hints of the delight the garage women took in their ambiguous sexual positioning survive in the gender-bending stories told by those who worked at the garage. Gabrielle Fleury, a Frenchwoman who had been severely burned while serving with the British Land Army in World War I, was remembered by a number of the women I interviewed as one of the most memorable of the garage workers. They recalled with amusement her short temper, florid language, and inability to memorize the streets of Melbourne as well as her powerful build and scarred body, which made people who did not know her assume she was a man. Using the surname address adopted in the garage, Jones, another long-term employee who was described as tall and thin and not very feminine looking, was remembered as the heroine of an apocryphal story of gender confusion. It told of a man who burst into the garage asking, "Is there anywhere a bloke can have a leak?"[17] Jones was said to have been completely unfazed and continued working on the car, while the poor man was greatly embarrassed for his indiscretion.

Stories of identity confusion were among the most commonly remembered and recounted. In them traffic police stopped the garage's chauffeurs, believing them to be under-age boys, conservative Melbourne society was shocked by their androgyny, unsuspecting men addressed the women in crude language, and bodily scarring could literally erase gender difference. There were recurring jokes about a garage "manned" entirely by women, and Alice Anderson was most often described as being like an energetic and charming boy. The dearth of terms with which to represent the pleasures of a specifically female technical proficiency suggests the difficulties that adhered to women's claim to a place on the workshop floor, but it also opened a space for jokes at the expense of men. The women took great pleasure in those jokes, delighting in the ways the garage could rattle the certainties of masculine entitlement to particular spaces, languages, comportments, and competencies. The jokes highlighted how the enterprise was not just a bid for meaningful work but also a public declaration of defiance. They gave voice to women's desires for change enacted at the everyday level of bodily actions, signs, and meanings.

These women's masculine style was an unstable ground of signification, however, over which the women had only limited control. In 1919 the progressive Melbourne press had interpreted their female masculinity as evidence of their professionalism and spirited independence. It was taken as an indication that they were advanced and admirably modern women. But during the second half of the 1920s the permissive era of gender experimentation drew to a close, and a new sexual conservatism came to dominate Australian society. Increasingly, the garage worker's female masculinity risked signaling a disreputable and deviant lifestyle, instead of forward-thinking modernity. Responding to that shifting ground of female modernity was a delicate task, which presented ongoing problems in the day-to-day life of the garage that only increased as the women grew older.

With her upper-class connections, lively personality, and charm, Alice Anderson was able to generate an enormous amount of goodwill and free publicity for her garage, right up until her death in 1926. Like the militant wings of the British, American, and French feminist movements, she adopted Joan of Arc as a sign of her earnestness. She wore a tiepin inspired by Joan of Arc on her uniform and inscribed it with the phrase "Qui n'a risque, rien n'a rien" (Nothing ventured, nothing gained). Joan of Arc, who was canonized in 1920, was a malleable symbol of female rebellion. She was courageous, fervently Christian, mobile, bob-cut, cross-dressed, sexually ambiguous, and single-mindedly prepared to risk all for a just cause. As a woman who had resisted the gender order from a position of feminine strength, Joan of Arc suggested an instantly recognizable and high-minded interpretation of the garage women's masculine activities.

Their attempts to present their actions in such elevated terms were only partially successful, however, and the pleasure the women took in creating their particular version of female masculinity sometimes threatened to overwhelm the business of the garage altogether. A magazine aimed at a sophisticated female readership headlined a feature article on Alice Anderson as "The Woman Who Does." Apparently an approving headline, it simultaneously resonated with the title of Grant Allen's scandalous New Woman novel of 1897, *The Woman Who Did*, in which the feminist heroine and an advocate for free love has a child out of wedlock and then dies tragically. Allen's novel was an example of a genre in which male writers pathologized feminist aspirations by linking them to sexual decadence, immorality, and emerging homosexual identities. So, while the banner "The Woman Who Does" placed over a feature article about Alice Anderson ostensibly celebrated her as a practical "do-er," a woman who put into action feminist precepts of sex equality, it also carried a hint of scandal, the whiff of a question. What else did Anderson and her "gallant little band" of khaki-clad mechanics and chauffeurs do?

To those in the know, whispers of lesbianism were never far away from the garage. In that broadly libertarian milieu a number of Alice Anderson's closest friends, as well as some of her staff and customers (the "University women," they were sometimes dubbed), were identified by her sisters as members of a lesbian network that centered on the garage. The owners of Lancewood private hospital, where she had been taken to die, were known as a lesbian couple, and Alice had often stayed with them in her hours away from the garage. But, while her sisters freely acknowledged that association, they were convinced that she attracted the sexual interest of women inadvertently. Alice "walked with the girls," as they put it, without being lesbian herself and was naive about the lesbians who were attracted to her and the business. Younger generations of the family are more inclined to conclude, however, that she was a lesbian.

Whatever Alice Anderson "did," or however she identified, there can be little doubt that she greatly enjoyed the transgressive elements of her professional and sexual positioning. A striking family photograph, carefully preserved because it was said to provide a perfect likeness of Alice's cheeky charm, shows her in chauffeur's uniform, sitting on a sand hill with an unknown woman of much more feminine appearance. The two are shown self-consciously playing their difference. Alice looks at her companion with a smile, holding her hand, an arm about her waist. It is a pose devoid of ironic comment at masculine pretensions, a parody that cross-dressing portraits frequently express with relish. Instead, it appears as a playful and flirtatious gesture, however uncertain in its wider import. Is Alice flirting with her apparent masculinity, even as she remains transparently female? Is she flirting with the young woman, who acknowledges the camera and smiles with the joke? Certainly, they are both flirting with the camera, the photographer, and the possibility of transgression. The caption on the back of the photo, supplied by one of Alice's sisters, directs an innocent interpretation: "Alice pretending to be the 'boyfriend' of a polio victim she had taken on an outing."

The photograph represents a moment that resists resolution. It is a fragment that reminds us of the joke as well as the seriousness of women's desires for the masculine license that automobility represented. For, in spite of Alice Anderson's buoyant personality and capacity for both social and physical mobility, apparently in contrast to her disabled client/friend, the position she occupied was fragile. She died too young, under circumstances that will always remain uncertain. Nor can we know how Alice Anderson might have continued her business had she lived longer. We can be sure, however, that the climate in which her garage operated was undergoing rapid change, making her business prospects doubtful. At the time that she died, the automobile service and repair sector was becoming increasingly regulated and institutionalized. Mechanical knowledge was formalized as a male, working-class trade that was unwelcoming to women. Technical improvements in automobile design meant that cars were more reliable, making it easier for people with no mechanical knowledge to be confident motorists. And, with growing car ownership in Australia, automobiles became an everyday technology in which owners opted to drive themselves rather than be chauffeured, so doing away with some of the functions that the Alice Anderson Motor Service provided. As they began to fall within the reach of less-wealthy families, an ethic of "driving" developed, rather than the earlier, more snooty practice of "motoring." Men from all social classes increasingly claimed automobiles as a quintessentially masculine technology, outside of the domain of even privileged women. These changes to the culture of automobility worked to make the style and rationale of the garage seem out of step with broader social changes by the end of the 1920s.

The business declined after Alice Anderson's death. It rarely rated a mention in the Melbourne press, and, though the driving school continued to flourish, the chauffeuring and mechanical work came to rely on a dwindling base of clients. The garage began to be viewed as an old-fashioned response to a version of automobile technology that no longer obtained. It became a curiosity in the local community, sometimes mocked, as the staff, clients, and equipment aged. Far from being considered modern and advanced, their uniforms were called "frumpy," and the women were easily dismissed as a remnant of an earlier era of class superiority, sexual oddity, and feminist pretension.

Alice Anderson and her garage have now been largely erased from Melbourne automobile histories, though in some social circles in eastern Melbourne, questions of whether Anderson was a lesbian who deliberately killed herself in the back of the garage continued to be asked well into the post–World War II years. The businessman who purchased her garage in the 1950s to expand his luxury car dealership summed up the elision when he wrote to me, "Alice Anderson Motors, was, I understand, a hire-car service operated by the two Anderson sisters prior to World War II and rumor has it that they committed suicide in the building."[18]

That bald version of events, a 1950s masculine narrative of female failure, works to write over and write out the fluidity and vitality of the gender experimentation of the early 1920s. It replaces the exuberance and creativity of female masculinity in that sexually progressive era with a revisionist story of women out of place in a man's world who must inevitably come to a doubly unpleasant end. In its misremembered version of literal sisterhood, however, the story inadvertently reinstates the atmosphere of lesbian sorority that once surrounded the Alice Anderson Motor Service. Over the twenty years of its existence her motor garage provided a group of young Australian women with pleasure, meaningful employment, and new sources of personal identity, while they worked to make less viable men's assumption of their exclusive claims to the power, mobility, and gratification that automobiles delivered.

CHAPTER 7

Driving Australian Modernity
Conquering Australia by Car

Australians first developed a fascination with driving a motorcar through the "dead heart" of the continent and "circum-motoring" its periphery during the 1920s. Although the first automobile trip through the center was made in the 1910s, a full decade after the earliest crossings of the United States, it was not until after World War I that transcontinental motorists emerged as national heroes in Australia. Popular mass media, including motoring magazines, newspapers, radio, and cinema newsreels, fed an interest in arduous automobile journeys through "trackless desert" sparsely settled by Europeans. The press paid particular attention to trips made by women and published accounts of their trials and triumphs as they "discovered Australia by car," "conquering" what was considered to be a male frontier. Photographs showed them kitted out in bush suits, Kodaks at the ready, automobiles fitted with long-range fuel tanks, water bags suspended from bumper bars, and camping supplies, spare parts, chains, and lengths of rattan matting roped to the running boards.

Unlike in the United States, where women's adventurous motoring was also an important element in the public representation of female motorists, the focus in Australia was not so much on simply crossing the continent but on driving around it, as close to the coastline as conditions allowed. As an island continent about the same size of the United States, the act of tracing its circumference was a gesture of integration, bringing the separate states and various physical environments into a symbolic unity. "Circum-motoring Australia," as it was often called, enacted a patrolling of the borders and confirmed white society's full possession of the continent. Women's transcontinental journeys in Australia also differed from those in the United States in that their trips did not attract significant sponsorship from automobile manufacturers. While carmakers and manufacturers of other motoring products seized the promotional opportunities that women's journeys provided, it was not commercial imperatives or stories of mass consumption that were at the forefront in Australia. Nor were women's automobile trips represented as a flight from the corrupting influences of congested cities, with their unwholesome tenements and their racial and ethnic conflicts, as was frequently the case in the United States at that time. Instead, press interest in the movement of female motorists through remote places in Australia was linked to the unfinished project of establishing a new nation in an old country. Arduous journeys had long been central to Australia's status as both a colonized and colonizing European outpost, and the public interest in women's automobile journeys through sparsely settled areas suggested that a new era had arrived—one in which white civilization prevailed and in which invasion had been replaced by peaceful settlement. Women's travel adventures were stories about a particularly Australian version of female modernity. They were about admirable women who would inherit that "new" country, at the same time as they worked to layer the landscape with tales of progress and white fortitude, helping to obscure the central dilemma of settler society in Australia—that the land belonged to someone else and the terms of its appropriation had not been acknowledged or resolved.

The widespread interest in early automobile adventurers was fueled by fantasies of a new kind of pioneering struggle, extended to people with little bush knowledge who were empowered by modern technologies to "break out new country" for white occupation. Significantly, it was only journeys made by city drivers that were considered remarkable, and arduous trips by residents of the outback were rarely noted as special achievements. Accounts by metropolitan drivers of the conditions they found and their success in conquering them were of greater interest in the South than the experiences of those who had made the tracks and traveled them as a matter of course. The fascination lay in urban automobile adventurers, who arrived with the express aim of passing through country as quickly as possible, charting routes for future tourists, and demonstrating the suitability of automobiles to Australian conditions. These travelers enabled Australians to view their country "with the eyes of the twentieth century," as one of the most famous of the overland expeditioners put it, returning with graphic images of Australia as a physically unified nation—securely under European domination, fully occupying its borders, and no longer needing to be troubled by the "wastes" that separated settlements clinging to the fringes of the continent.[1]

The rigors of travel, especially in the northern and central parts of the continent, lent themselves to heroic stories. Early motorists followed unsealed tracks and sometimes forged their own trails in immense heat, across long stretches of country, where the only reliable water was bore water at government stock wells. Motorists laid lengths of rattan matting in front of their wheels to cross stretches of sandy desert. If they were caught by rains, they could spend days digging their cars out of black soil bogs. They pushed and winched cars across flooded rivers and drove over railway trestles when no other route was possible. In high grass country the vegetation could be taller than their cars, concealing washouts, potholes, and giant anthills solid enough to split an oil pan. Grass fires, endless tire punctures, running short of fuel and water, illness, accidents, and breakdowns far from help were serious dangers. No matter what the hardship, however, press articles about women travelers in the "never-never" reassured readers that they were far from being daunting "Amazons." Journalists invariably emphasized their femininity, their blithe confidence in their automobiles, and their "thrilling encounters" with "wild blacks" in the "back blocks" of the nation.

Even more than men, female transcontinental motorists were received as heroes in the towns and cities they passed through. Motoring interests promoted their journeys by feeding information to the press in advance of their progress and organizing receptions for them. Crowds turned out to welcome and farewell them, foreshadowing the adulation that transcontinental aviators such as Amy Johnson and Jean Batten experienced a few years later in even greater measure. Their cars were decorated with autographs, streamers, and wildflowers. They were entertained by car clubs, presented with gifts, welcomed by city dignitaries, interviewed by the press, and filmed for cinema newsreels. Some of these cross-country drivers recorded radio talks along the way and delivered lanternslide lectures in theaters and cinemas after they returned, appearing onstage as part of the variety bill. The entertainment pages of daily newspapers framed their illustrated talks as pure action series plot. One review, entitled "Two in a Car: Women Wanderers in the Waste," described two ladies "who had gone on adventure to prove that women were equal to men when it came to traveling this little known country." They had to overhaul their engine far from any help and overcome constant danger from the waterless spaces, snakes, grass eight feet high, violent thunderstorms that turned black soil plains into bogs, and bushfires that threatened to engulf their car loaded with gasoline.[2]

These women did not describe their travels, however, as always teetering on the edge of disaster. As they pushed their cars into territory barely accessible by car, they wrote and spoke quite differently about their reasons for undertaking their journeys. Indeed, Marion Bell, who in 1926 was the first woman to drive around

Australia, went to a great deal of trouble to contradict highly dramatized press accounts of her travel. When the *Adelaide Advertiser* reported that she had driven her car between warring Aboriginal tribes at Roy Hill Station in the Northern Territory and that "her confident bearing and offer of sweets resulted in a suspension of hostilities," she protested against the sensational accounts. She "emphatically denied the embellished version of the happening and indignantly inveighed against the newspapers, which by publishing the ridiculous tale assisted in spreading a false atmosphere concerning her and her trip."[3] When we are able to get a sense of the women's own understanding of their actions, it reveals a different emphasis from the overblown descriptions put forward by male commentators. Women's accounts of their "overlanding performances," as they were dubbed in some manufacturers' advertisements, offer an alternate expression of what it meant to be a modern woman in a colonial society.

The year for Circum-motoring Australia was 1925, when three parties departed from Perth in Western Australia, hoping to be the first to make the circuit around the edge of the continent. Marion Bell and her eleven-year-old daughter, also called Marion Bell, left in October. They were the third to leave Perth and third to make it back, taking almost six months to complete the circuit. Photographs of their departure in an Oldsmobile Six, a medium-sized car at the cheaper end of the market, show a dashing Mrs. Bell with bobbed hair, carefully outfitted as the intrepid motorist in a leather jacket, corduroy breeches, high leather boots, and a "Tom Mix" hat, the cowboy style popularized by the American movie star. Her daughter was dressed much more conventionally, in a white dress and ankle sox. Young Marion was brought along for company and respectability, and it was her duty to record the speedometer readings, fill the radiator, keep plenty of air in the tires, and open hundreds of gates blocking their path — they counted thirty-seven in one day alone.

The mother and daughter headed north from Perth in mid-October 1925, through tiny settlements and cattle stations along the northwestern coast and then on to the top end, tackling the hardest and least-traveled sectors before the wet season made driving impossible. It took almost eight weeks before they reached the comparative comfort of the northeastern coastline, where they rested and enjoyed the attention of admiring crowds and newspaper reporters. Reaching Brisbane, Sydney, Melbourne, then Adelaide was a triumph for Mrs. Bell and her daughter. They were entertained by prominent motorists, honored with special functions at automobile clubs, and invited to racing meets and air shows. Their car, still bearing the scars of its arduous journey, was put on display in Oldsmobile showrooms in major cities, while they rested and socialized. When they returned to Perth in

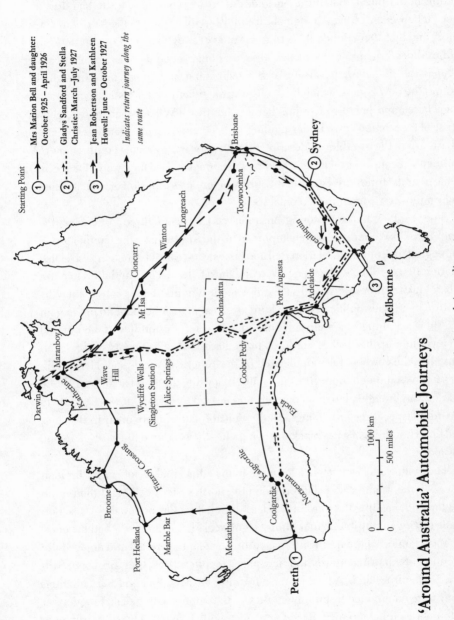

'Around Australia' Automobile Journeys

Circling Australia by car. Map of the earliest women's Around Australia motor journeys, 1925–27. Drawn by Peter Johnson.

early April, they received the biggest reception of all. Thousands of people came to the main square to welcome them home. There were cinematographers, photographers, welcoming speeches, and a crowded press conference.

When asked what had prompted her to make the trip, Mrs. Bell expressed an appetite for novelty that placed her at odds with notions of domestic femininity. "I have always been inclined that way," she laughingly explained, "always inclined to be a wanderer. I could not stay at any place for more than two years without becoming utterly sick of it, and obsessed by a desire for change."[4] Newspaper stories of Marion Bell's journey hinted at the discomfort she caused, unsettling certainties about women's proper place. Even as they praised her courage, few press articles were free from the assumption that she had naively and carelessly trespassed into masculine space, thereby creating problems for those around her. There were suggestions that it was mainly good luck and the kindness of men that brought the mother and daughter team through their journey unscathed, and shortly after they returned to Perth some of the earlier praise and admiration for her turned sour. Acrimony between Mrs. Bell and the second man to complete the circuit, Joshua Warner, was played out in their local newspaper and soon picked up by national motoring magazines and the tabloid press. Both drivers claimed the other was not competent to make the journey and would not have made it all the way without help.

As Joshua Warner divulged to the motoring reporter of the *Sydney Mail*, he was only three days out of Perth in his small four-cylinder Citroën when Mrs. Bell and her young daughter, Marion, overtook him. "You can imagine my surprise on finding a woman and a young child driving out into the 'blue' without a care in the world," he told the reporter, "in vain did I try to persuade her to return."[5] Marion Bell, according to Joshua Warner, spiritedly defended her ability to continue the journey. "Mrs. Bell remarked that her six-cylinder car could go anywhere a four-cylinder car could, and she could do what any white man could do."[6] And so the dispute between them grew, with Marion Bell vehemently contesting the privileges of the road that white male travelers took for granted. Warner declared that Mrs. Bell had left with no plans and no provisions. He had been forced to provide her with fuel from his own limited stocks and had been delayed by continually digging her car out of bogs.

Marion Bell strongly disputed his version of events and indignantly denied needing his help, declaring instead it was she who had assisted Warner to get his car through. At no time did he have to tow her out of difficulties; rather, she had to help him on many occasions. She had to loan oil to Mr. Warner continually, and, because he did not even know how to change the wheel of his Citroën car, she had to do it for him. "For hundreds of miles I led Mr. Warner, my car beating down the bush, and forming a track for the light Citroën," she proclaimed.[7] Marion Bell's proud declarations of independence, that she could "do what any white man could do," indicates that being recognized for getting around the continent by her own efforts constituted the value of the journey for her. She often asserted that no one else had touched a wheel of her car since she had left home, not even in the worst bogs. She had driven the car across every river herself, sometimes being the first to make the crossing by car.[8] She stoutly maintained her feisty approach to the criticism that she had blundered into territory that was only barely safe for white men and to accusations that she had not returned in the car that she had set out from Perth in, with a combativeness that was characteristic of her life. A proficient bush woman, having grown up in remote parts of New Zealand, she had run away to Melbourne as the young mistress of a prominent New Zealand barrister, with whom she had a son. After his death she married Norman Bell and settled in Perth. By 1923 she owned a bus company, Marion Bell's Char-a-Banc Service, which operated six buses in the tough and unregulated climate of early motor transportation in that city. She continued to run a motor garage, taxi business, and ambulance service in Perth until well after World War II and had painted in large letters across the display window of her garage, MARION BELL: THE FIRST WOMAN TO DRIVE AROUND AUSTRALIA.

Less than a year after Marion Bell and her daughter completed their controversial circuit of the continent, two New Zealand women, Gladys Sandford and Stella Christie, set out to drive around Australia in a Hudson Essex Six, one of the first closed cars to be brought into Australia. Just as it had been for Marion Bell, getting through on their own efforts was their foremost challenge and greatest point of pride. Essex published a full-page advertisement written by Gladys Sandford and illustrated with her photographs of their car in the extreme conditions. Characteristically, her testimonial brimmed with confidence, declaring that her interest in cars was thoroughly practical. "I am conversant with the technical side of cars and am able to do ordinary running repairs and overhauls myself," she wrote.[9]

Sandford and Christie had set off from Sydney a year after Marion Bell, heading south through Melbourne and Adelaide. When they reached Perth, record rains prevented them from traveling farther around the coast as they had planned, and they were forced to backtrack to Adelaide and then drive northward through the center of the continent to Darwin. Floods again stopped them from driving east to Queensland, so they returned to Sydney via Alice Springs, Adelaide, and

Melbourne, having traveled more than eleven thousand miles in four months. The press turned their altered plans to good account with headlines such as "Australia Criss-Crossed by Car"; "Four Times across Australia!"; "Pluck, Petrol and Petticoats: The 'Do and Dare' of Two NZ Women"; and "Men Superfluous: Sydney-Perth Car Trip: Men Not Wanted." Enthusiastic crowds met the women in all the major centers they passed through, and back home in New Zealand they toured the variety theater circuit in their driving outfits of breeches, boots, and puttees, presenting "magic lantern" lectures of their trip.

Like Marion Bell, Gladys Sandford had traveled a great deal and learned to look after herself, though in Sandford's case it was World War I and not sexual scandal that had propelled her out of New Zealand. She had been born in Australia and moved to New Zealand as a child. Early in the war she followed her first husband and two brothers to the European front, where all three were killed—her husband only weeks before the armistice was declared. Sandford worked as a driver with the New Zealand Ambulance Corps in Egypt and France. She stayed on in England after the war, organizing the Motor Transport Division at the New Zealand military hospital on the outskirts of London, and was eventually discharged with the honor of being the only New Zealand woman awarded a regimental number—something about which she was always proud. She married John Sandford, an Australian airman, in 1920 and spent some years in India and Egypt, where she accompanied him on diplomatic postings. When the marriage failed, in 1924, Gladys Sandford returned to New Zealand, supporting herself as a car saleswoman. She used her military connections in 1925 to learn how to fly, and by early the next year she was advertising for a female companion to join her on an around Australia drive. It was the first of several adventures she was planning, she told the press. The Australian trip was to be a rehearsal for a Cape-to-Cairo "jaunt," and she had plans to be the first woman to fly across the Tasman Sea, from New Zealand to Australia.

Her companion on the trip, Stella Christie, was the daughter of one of the directors of the firm that had loaned the car. Christie, who was much younger than her and not able to drive, was a great disappointment to Sandford. The young woman was awkward, shy, unassuming, and barely mentioned in press reports. Photographs show her, small and retiring, beside the strong and impressive Sandford, who was always described in glowing terms. "A splendid type of woman," according to the *Northern Territory Times*, "with a pair of glorious blue eyes and fresh fair complexion, quite undamaged by the long jaunt."[10]

Both Marion Bell and Gladys Sandford kept journals of their trip. Bell's were sketchy and erratic notes, made with the intention of writing a book on her travels, while Sandford's journal was written as a continuous narrative, typed and smoothed by her numerous theater and radio appearances after the event. Neither journal was published as a book, but each formed the basis of extensive press articles published after they returned.[11] What their journals had in common was that the women only minimally focused on the landscape they passed through, except when high grass, sand dunes, rivers, and bogs presented obstacles to their movement or challenges to their progress. Rather, it was the travelers themselves and the impact they made on those they met that constituted the major subject of their writing. Both women framed themselves as spectacles, rather than focusing on the ostensible objects of their curiosity—the landscape they passed through and the people they encountered.

Marion Bell's journal noted the many welcomes that she and her daughter received at bush hotels and at cattle, telegraph, and police stations along the way. She described the meals and cups of tea offered to them, recorded admiring comments and the way people waited by the side of the road, sometimes for days, to be able to chat with them. Cameras flashed upon their arrival, new friends drove with them to see them to the right road, and their autograph books were inscribed with extravagant testimonials to their charm and courage. In her diary Marion Bell presented herself as a competent and courageous woman, always made welcome in the "top end" by men who appreciated the difficulties of her journey and had been starving for the company of white women. She told her local newspaper that homesteaders in the Northern Territory called her "the best woman sport in the world" and had made an application to the government to have one of the local ranges named the "Marion Range."[12]

Gladys Sandford was even more careful to recount the admiring and occasionally humorous comments of strangers whom she and Stella Christie met. She proudly noted that Aboriginal women on the Nullarbor Plain, who had been assigned by station owners to wash the travelers' clothes, refused to believe she was a woman and called her "mamaloo," which she translated as "man." She recorded with amusement that one station owner who came out to look for them "tells us that he expected to find us starving and in hysterics, with our skirts dragging in the mud." Their impact at a small town post office, where a boy in his mother's arms declared her to be a man and not a lady, she noted with humor. And she carefully documented the comments of a man who read the inscription on the side of the Essex, SYDNEY TO PERTH—THEY'VE UNDERTAKEN A LONG TRIP. On seeing the additional word, DARWIN, a distant city to the north, he muttered, "Hell, they've got pluck."[13]

Gladys Sandford documented the intense physicality of the journey in even greater detail, describing with care the pleasures and agonies of hard work in high temperatures. She wrote of becoming covered in grease, mud, and flies and recorded with pride her offer to repair broken springs for men whose cars had been stuck in mud, reveling in joining them, slipping and sliding through the slush to push their cars out. Sandford recounted at length the work involved in repairing the clutch plate "in the heat of the day and with no pit."[14] She described her groans as she struggled to lift the gearbox and carry it to the top of a bank and out of the sand where the car had become stuck. The two women swore and cursed as they wrestled the new corks into the clutch plates in the blazing heat, Sandford compressing the plates between her knees with a tire lever and hammer handle, both working desperately to secure the nuts and bolts that would hold the assemblage together.

With a similar relish in the intense physicality of the enterprise, Sandford elaborated on the frustration of getting out of "a truly terrifying place," Bridge Creek south of Darwin, where they were warned it would be impossible to take a car through. Sandford and Christie spent thirty hours building up the banks of the creek with trees they had felled with a tomahawk and constructing a Spanish windlass, with the help of a passing railway foreman, so they could winch the car to the other side with fencing wire. She recorded breaking into a "haka," a Maori dance of aggression and triumph, after they finally succeeded. Marion Bell similarly took care to describe her pleasure in the physical challenge of the work. She recorded in her diary that she could not pass by a "poor driver" who had bogged his truck near Maranboy in Central Australia. She waded into the mud barefoot, helped him jack up the wheels so that they could pack up the bog with wood, and then towed the truck out with her Oldsmobile.[15]

The relish the women took in the physicality of the work was glossed over in journalists' reports. Instead, male reporters invariably emphasized that, for all their apparent "grit and determination," the women remained reassuringly feminine — always slight, attractive, and charming. The convoluted prose used to describe their appearance indicates the difficulty journalists found in accommodating the simultaneous possibility of their femininity and competence. One of them declared approvingly that Marion Bell was not a woman of "Amazonian stature with a set of muscles in keeping with these somewhat Herculean traditions," nor was she "the masculine, mannish-clad type." "She seems more the coupe twenty-mile tour type," another journalist expressed with relief. Her stamina and resiliency, he decided, lay in the "wonderful nerves" that had seen her through.[16]

The apparent disjunction between the women's femininity and the courage, stamina, and vitality their actions so clearly demonstrated was a point that received much attention. Accounts regularly emphasized that they were in no way "abnormal" women and that modern, mannish activities need not lead to mannish modern women. In these reassurances the work involved in *appearing* as independent female travelers—whether it was defiantly, in "male attire" and greasy breeches, or whether it was modestly, in proper feminine style—was hidden from view. In the eastern cities where they were not known, Marion Bell and her daughter were photographed in dresses, waving cheerfully for the camera, garlanded with flowers, and cuddling an orphaned lamb they had found on the road. Marion Bell revealed in her journal that they changed from khaki overalls into the dresses they had carefully stored in dust-proof bags and brought out just before they entered a town. In Perth, where she was well known as a bus driver, she asked her husband to meet her with her leather driving outfit of coat, breeches, and cap. Gladys Sandford and Stella Christie, on the other hand, always remained in breeches and shirts for their city receptions, though Sandford's journal indicated that it made them feel somewhat self-conscious. For Sandford, who maintained strong connections to ex-servicemen's organizations throughout her life, their outfits were a proud reference to her motoring competence and to her exemplary military service.

How women responded to the scrutiny they experienced was part of a gendered discourse of the female body, in which the ways they chose to appear as well as their comportment on the road were speaking acts in their own right. For male journalists the central issue lay in how their femininity should be placed against the evident competence of their actions, a competence that seemed to compromise those supposedly natural qualities that made them women. The women, on the other hand, had their own sets of emphases on their bodies and sought to assert control over the terms in which they were represented. They tried to tell a story of energetic female embodiment in which neither sexuality nor female weakness was placed in the foreground. In celebrating the intense physicality of such extreme forms of travel in terms of a specifically female technical competence, or at least into a technical competence in which their female bodies could properly partake, they were putting forward a different kind of female body, one that could safely travel in remote areas. It was not a dependent, sexualized object of male appraisal but a whole body—absorbed, effective, diffusely skilled, and purposively oriented to the world of things. Viewed in this way, the women motorists were answering the pervasive sexualization of the female body in those postwar years with a different set of descriptive terms. Their "overlanding performances," in a time-honored tactic within feminist traditions, was a particular kind of performance art, in which the ways they appeared and their self-descriptions constituted a bodily reinscription that answered to the sexualizing of their bodies that they were attempting to sidestep—indeed, that they needed to sidestep in order to travel safely.

Jean Robertson and Kathleen Howell, the third pair of long-distance motorists, represented quite a different style of female adventurous motoring. Much less confrontational and abrasive than Marion Bell and not as emphatically competent as Gladys Sandford, they were young society women—good sports and pals to the men with whom they competed in automobile club events. The women had been school friends and from 1925 motoring partners with considerable success in reliability trials and rallies organized by the Automobile Club of Victoria. Jean Robertson had learned mechanical repair work as a pupil at the Alice Anderson Motor Service after she left school. Press photographs show them as dashing and stylish in their powerful Lancia Lambda, a low-slung, Italian touring car that was priced at about one thousand pounds, well beyond Marion Bell or Gladys Sandford's means. When they raced the transcontinental train from Perth to Adelaide, driving seventeen hundred miles nonstop for two days and ten hours, the press described them as "two young slips of girls," "garbed in riding trousers, silk shirts and slouch hats."[17] Their driving partnership continued into the Depression years, and, in their most remarkable journey of all, they were members of an Australian party who drove three light cars from Melbourne to Palermo to compete in the Monte Carlo Rally of 1932.

Their photograph album documents a journey north through Alice Springs to Darwin at the same time that Gladys Sandford and Stella Christie were traveling south on the same route. It shows them inching across sand hills, bumping across miles of gibber stone (or boulder-strewn) desert, crossing swollen rivers by driving their car over high railway trestles, placing additional sleepers under the tracks to support the wheels, and repairing countless punctures. They logged their mileages at the request of their gasoline sponsor, and the figures were included in Shell's first blueprint strip maps of the Central Desert. The women's album documents the emergence of motor tourism through remote Australia as a privileged and pleasurable activity. The women sip tea on homestead verandahs, stylishly outfitted in riding breeches, boots, shirts, ties, cloche hats, and long gabardine coats. There are photographs of magnetic anthills, water holes, young policemen on motorbikes, telegraph stations, itinerant "swaggies" they encountered on the road, "alligators," and views of Darwin's Chinatown. Aboriginal women and children are lined up for their camera, and Aboriginal men are photographed chained together as pris-

Kathleen Howell with Barney the dog. Taking care of sponsors in Central Australia, 1927. Courtesy of the La Trobe Picture Collection, State Library of Victoria.

oners, their crimes unnamed and not questioned. Their album also contains photographs of two women at Singleton Station, 240 miles north of Alice Springs, their immobilized car shaded by a "whirlie," or brush shelter. The women were Gladys Sandford and Stella Christie, who had become stranded in desert country with a damaged gearbox housing, dangerously short of food and water. With no passersby in four days, they had saved themselves by walking into Singleton Station and telegraphing for replacement parts. Robertson and Howell found them camping in their Essex, waiting for the gearbox that, coincidentally, they were carrying north from Alice Springs.

Singleton Station, leased by Mr. and Mrs. Crook and their two daughters, was a painfully poor and marginal enterprise seven miles from Wycliffe Wells. The family had emigrated from Kent to the Central Desert almost twenty years earlier. Photographs showed them to be lean outback "battlers," their plain clothing frozen in a turn-of-the-century style—the father aging, the mother frail and unwell, and the daughters awkward and shy. The Crook sisters were becoming regular figures in stories of outback isolation, as transcontinental travelers returned with tales of two unmarried white girls in their early twenties who had lived in the center of Australia all their lives and had never seen a white baby or a train. They drew a great deal of attention from the fact that they lived such a hard life, living in a

Putting out the "hall runners" to pass through a dry creek bed, Central Australia, 1927. Courtesy of La Trobe Picture Collection, State Library of Victoria.

brush shelter, working cattle with their father, and pulling water from the government well for the mobs of cattle that traveled the stock route, a world away from the frivolous concerns of their "flapper" sisters in the city and certainly not at all like the romantic "bush girl" heroines of the Australian fiction or silent screen. Gladys Sandford portrayed the sisters as hungry for contact with the outside world, looking to the younger of the visitors to induct them into modern, girlish interests. They became Stella Christie's "shadow," and for the first time Christie was given her moment in the journal, her youth unexpectedly an asset. "Did she believe in love at first sight?" the Crook sisters wanted to know, and in the evenings they sat on gasoline cases under a desert sky learning the latest American songs from their visitors, "Yes Sir, That's My Baby" and "I'm Sitting on Top of the World." Gladys Sandford, with little interest in fantasies of romantic love, passed the three weeks capturing finches in the grass seed catcher that protected the Essex radiator and learning to roll her own cigarettes. As soon as the gearbox was delivered, she set to fitting it, and the two parties of women continued their journeys, each heading in opposite directions.

Kathleen Howell's view of their meeting was not only recorded by her camera but was later published in a Melbourne motoring magazine. That account, spare and matter-of-fact, was not at all in the epic tone of Gladys Sandford's journal and was in striking contrast to the overblown style in which women's journeys were often reported in the press. She calculated their load, the gas they needed to cover the 750 miles from Alice Springs to Katherine, noted that their radiator was blocked by the minerals from the bore water, and the difficulty they had in finding room for the clutch and gearbox they had been asked to carry for Gladys Sandford at Singleton Station. "It was rather unique," she wrote, "that the two parties of women motorists should meet as they did three hundred miles from Alice Springs, as near to the center of Australia as we could be."[18] The Crooks, she thought, had never had such a visitation of white people before, and Sandford and Christie's breakdown in that spot was a real godsend to them.

The significance of that desert encounter between seven white women, "as near to the center of Australia as we could be," reached far beyond the new kinds of femininity performed in adventurous motoring at that time. The depth of public interest in the women's trips derived not simply from a fascination with expressions of courageous colonial femininity but also from the ways that the women's journeys tapped into unsettling themes at the heart of national life. Looking back, women motorists' actions touched on misgivings about the foundations of Australian settlement, and, looking toward the future, they raised questions about the part that white women might play in shaping the modern nation. The meaning of that brief encounter between the seven women—four of them passing through and three of them settlers who remained behind—can best be understood in terms of the historical and emotional proximity of colonization in those remote areas. The independent mobility of the four city women promised a new kind of mastery over the country, where a deep knowledge of it was no longer a prerequisite for survival and a brief visit was enough to confer a sense of belonging. The isolation of the three settler women, on the other hand, and the meager living they earned from the desert hinted at concerns about the future of white occupation in Central Australia. The Crooks' poverty only barely differentiated them from their Aboriginal neighbors, their rough living conditions placing their white civility at risk. The family's scant survival raised questions about who could rightly claim to own that country and what relations should properly exist between new settlers and original inhabitants. For, in noting the significance of a temporary influx of white women, bringing pleasure and relief to the women of the Crook family, there was an implicit acknowledgment that Singleton Station was deep in someone else's country and that the Crooks' usual companions were precisely those people who were being dispossessed by white settlement and tourist travel.

That closeness of colonization ensured that relations between the women travelers and Aboriginal people were central to the stories told about their trips. Marion Bell and Gladys Sandford spoke of being repeatedly warned against "wild blacks" when they announced their travel plans and were urged to carry revolvers for protection. Journalists always questioned female travelers closely about their experiences with Aborigines, and stories of their contact—which always assumed that it was the white women who were exposed to danger—exerted a great fascination in the cities. Upon their return the women invariably made a point of declaring that such fears were completely unfounded. As Gladys Sandford put it, she was constantly advised to carry a revolver at the hip and endeavor to make it to stations by nightfall when she was north of Alice Springs. That concern only showed how little most of the Australian population knew of the conditions, she said. "We were never once troubled by natives and would make camps at night on the track without ever any fear."[19] Marion Bell similarly emphasized her good relationships with Aboriginal people. Perth newspapers reported how everyone was amazed that she had traveled around the country safely, as white people were complaining of "trouble with the blacks." Mrs. Bell stated, however, that "she [had] experienced no difficulty—in fact she [had] often been much assisted by them."[20]

Such declarations of white women's sense of security as they traveled through Aboriginal country suggest how the myths of race and gender are interdependent. If Aborigines were dangerous savages, as popular opinion maintained, then white women must be vulnerable and require the protection of white men. Conversely, if white women's power—courtesy rather than firearms—was sufficient to ensure their safety, as women travelers' frequently declared, then Aborigines must respond to it, so establishing that they too must be courteous. Thus, in the context of these women's stories the validation of their power and the possibility of their independent travel affirmed a degree of Aboriginal civility that the standard myths of the outback—steeped in stories of white frontier masculinity, as they were—found difficult to incorporate.

Race relations at that time in both the northern and central parts of the country continued to bear the stamp of frontier violence characteristic of the earlier period of colonization to the south. Newspaper reports and parliamentary inquiries published in the 1920s confirmed that massacres of Indigenous people continued—the Forrest River massacre in the Kimberleys in 1926 and the Coniston massacre in Central Australia in 1928—and a number of advocacy groups, such as the Aboriginal Protection League, had been formed in the southern states to organize the small but growing dissatisfaction with the policies being pursued in the North. Some white feminists focused their activism on the rights of women and children, producing critiques of black and white frontier masculinity. They argued that Aboriginal women and children required protection not only from their "savage" male kinfolk, as had long been asserted, but also from the sexual exploitation of white men. Frontier settlers had long assumed the right of sexual access to Aboriginal women, and road and rail transport were bringing even more white men to remote areas. Evidence of sexual contact was visible as the highly charged "half-caste problem"—the increasing numbers of children that white men did not acknowledge and who were the targets of brutal practices of child removal by the state, continuing into the early 1970s.

Neither the motoring press nor the women motorists themselves hinted at any critique of frontier violence or of white male sexuality. Motoring interests emphasized the commercial implications of their trips, arguing that the women's success in getting through safely was a guarantee that the whole country was securely under white domination and that stable settlement now held sway. The fact that Aborigines, sometimes as chained prisoners, could be forced to haul cars across rivers and to build the very roads those travelers used was rarely mentioned in newspaper stories. Instead, Aborigines were portrayed as a tourist resource, part of an exotic landscape where encounters between white travelers and Aboriginal people could be reported as "thrilling experiences" in which tourists "saw the blackfellow in his native habitat." The stories worked to secure three important preconditions for a white Australian identity. It consigned Aborigines to a past into which they were doomed to fade quietly away, unable to share in a contemporary present. It provided a screen on which to project the superiority of European technology, legitimizing and making inevitable the transfer of the land from Aboriginal people to white ownership. And, finally, it worked to "historicize" the landscape in European terms, endowing the country with new stories of white heroism and adventure.

The view of automobiles as a technology that served to draw a sharp line between advanced and primitive people, casting whites as the only legitimate inheritors of the future, was expressed in many of the advertisements for motoring products that drew upon women's around-Australia trips. In one such image an Aboriginal man armed with traditional weapons watches a map of Australia from a distance, observing his country now encircled by Mrs. Bell's car tracks and bowing his head in apparent homage to her "Triumphant Return to Perth." In another an Aboriginal warrior is drawn fleeing across the desert, away from the adventurous mother and daughter and their superior technology. Such advertisements helped to consolidate white claims over the whole of the continent and worked to legitimate the displacement of Aborigines, apparently with their voluntary acquiescence. In contrast to the triumph of the white women motorists, who were located as women of the future, Aboriginal men were cast as part of the natural landscape, intriguing, picturesque, evidence of an ancient history to which a young nation could proudly make claim but that had now been inevitably displaced.

As for the women motorists themselves, in their desire to be positioned alongside white men in narratives of modern nation building, they were disinclined to criticize frontier masculinity or to question the race relations on which the settlement of northern Australia had come to depend. Marion Bell's reported declaration to Joshua Warner at the beginning of their journey that "she could do what any white man could do" was an acknowledgment of prior ownership of the territory she was passing through, but it also indicates that she viewed her mobile competence as a means of aligning herself with white men as the natural inheritors of that country. Gladys Sandford even more explicitly used her identification with a modern technology to position herself as quite distinct from "primitive" peoples, entitled to intervene in their lives as she wished. One of her favorite stories, retold many times in her journal and in her radio and theater performances, related her experiences at a corroboree, or theatrical dance performance, arranged by the manager of Maranboy Aboriginal Reserve for his visitors' entertainment. Maranboy was a tin mine fifty miles south of Katherine in the Northern Territory, which relied almost entirely on Aboriginal labor. On noticing two newborn babies in the group of "myalls," or Aborigines who had little contact with whites, Sandford asked the parents through an interpreter if the babies had names. Believing the parents had said they had not been named, she decided to "christen" them herself. For the boy she pronounced, "This fella him name Hudson," and for the girl, "This fella him name Essex." She noted that the people had great difficulty pronouncing the x and joked that the child would always be called "Essek."[21]

In presuming the right to name and then choosing to give Aboriginal babies the name of her prized automobile, Gladys Sandford was imagining herself as a player in the project of modernizing Central Australia. Wild and potentially dangerous

Technological power confirms white possession of the land. Mrs. Marion Bell and her daughter reach Adelaide. *South Australian Register*, 19 February 1926, 16. By permission of the National Library of Australia.

Aborigines were brought into her modern, Christian orbit by her power to name. They could only be passive or uncomprehending in that exchange and, humorously, get it wrong, while modern white women—competent, independent, courageous, and mobile—could move through the landscape, breaking new ground not only for themselves and their sex but also for the nation.

Women's motoring activities worked in particular ways within Australian national debates of the time, very differently from in Britain and the United States, and remind us of the disparate sources that shaped the nation that Australia was coming to be. Adventurous women motorists not only personified a new femininity, one that expanded social possibilities for settler women, but they also contributed to a project of creating foundational stories for an emerging nation. Their actions revitalized old narratives of white possession of the land legitimized through technological superiority, for a younger generation and for a new era. Images of modern white women, now able to join men in their pioneering work, stepping forward to become equal partners in "taming the country" ran through all of the stories and animated the claims of various interest groups, including the women themselves. Their trips were welcomed as a demonstration of how automobile technology enabled a new generation of settler women to inscribe white culture onto the bush. The stories that were told about them illustrated how women motorists' ambitions and actions were part of bringing into being a new nation, one in which technologies would make viable the white habitation of remote areas and in which women's resourcefulness and courage offered another register to make the ongoing transfer of land from Aboriginal possession into the hands of white settlers seem entirely proper and natural.

CHAPTER 8

Machines as the Measure of Women
Cape-to-Cairo by Automobile

Women's transcontinental journeys in Australia during the 1920s were given meaning within the project of creating a modern, unified nation. Motorcars, as objects that shaped both personal and national identities, enabled particular women to insert themselves into the practices and conversations of nation building that were so central to Australian public life at that time. Through their actions and words transcontinental women motorists found an honorable place for themselves within the big debates about how the new nation might fully occupy its borders. Western women's automobile journeys sometimes went beyond their respective country's borders, however, occupying a transnational, colonial setting. Women's stories of colonial travel, in contrast to the masculine narratives that dominate the field, offer an alternative perspective on the links between technological progress, gendered identities, national loyalties, and imperial ideologies at that historical moment, for the ways in which women's travel narratives were produced and received were not the same as those of male travelers. Women traveled differently, they wrote differently about how they traveled, and their audiences read the stories they told differently. Their narratives illustrate how the colonial context within which their travel occurred was not just a neutral backdrop to accounts of Western supremacy and technological progress. Instead, they show how the colonial setting operated as an active agent in stories of early automobile travel, in which relations between colonized and colonizers were continually being remade and the ripping yarns of colonial adventuring spoke of disquiet and uncertainty, as much as they trumpeted superiority and a sense of white entitlement.

In 1930 two South African women of English descent, Margaret Belcher and

Ellen Budgell, took an automobile journey through the eastern side of the African continent. They drove an aging, secondhand Morris Oxford from their home in Cape Town, along the length of the continent to Cairo, and then via Paris and London back to the factory in Cowley, near Oxford, where their car had been manufactured. Their mode of travel was crucial to the story the women told. Automobile travel gave them a measure of autonomy from African labor, quite different from the overland Cape-to-Cairo treks made some twenty years earlier by British travelers such as Mary Hall or the American adventurer May French-Sheldon, who were carried along the route in shaded litters by teams of porters. Belcher and Budgell's account is an alternate story from the margins of the colonial project and reveals some of the complexities of a female positioning, in which women were obliged to inhabit a minor register within a masculine material practice and narrative genre.[1]

Even more than in the metropolitan centers, early automobiles were items of privileged personal consumption in the colonial setting. They were material objects whose very nuts and bolts loudly proclaimed the inequalities of global power relations. As an editorial in the *Motor Age* of Chicago in 1916 reminded its well-to-do readers, "Nearly every country around the globe has been obliged to give up its treasures" to produce the "cosmopolitan mechanism" of a motorcar. "Thousands of people from every clime, of every color," it continued, "have sweated and labored so that you might have this gilded plaything." For the lucky few it was a vehicle of pleasure, incorporating the "romance of the South Seas, the wild free air of the Texas prairies, the drudgery of underground toiling, the roar of the blast furnaces, the crack of the slave-drivers' whips over bare backs on the rubber plantations, the negro melody in cotton fields, the white heat of electric retorts, the frantic search for precious metals, the frenzied speculation of the oil booms and the tragedy of Siberia."[2]

Accompanying the editorial was a map, which placed a touring car hovering over the United States, the center of the automobile world. Component materials scattered across the globe—copper, wool, mica, chromium, mahogany, and vanadium—were shown being drawn to that nodal point, a knot of linkages congealed to form the American car. Through connections that had become so ordinary as to be hardly noticed, the article declared, an automobile was an object in which inequalities were frozen, and the hard labor of suffering "others" was materialized for the pleasure of the American consumer.

Not only did the editorial point to a careless certainty in the rightness of such privilege over colored and sweated peoples in every corner of the globe, including within the United States itself, but it also implied an American superiority

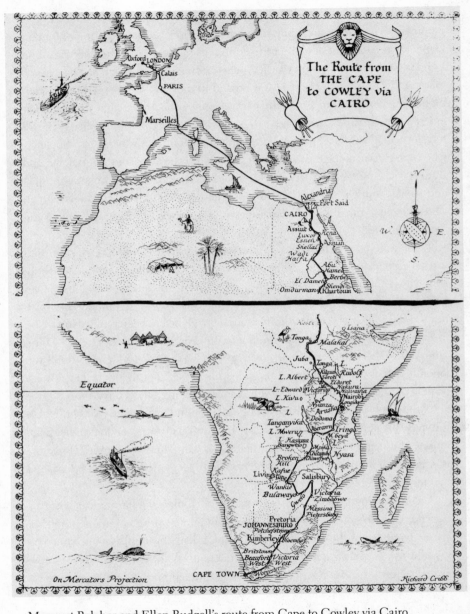

Margaret Belcher and Ellen Budgell's route from Cape to Cowley via Cairo in "Bohunkus," 1930. From M. L. Belcher, *Cape to Cowley via Cairo in a Light Car* (London: Methuen, 1932).

over the old centers of industrial power—the very European countries in which the automobile had first been developed. With European and British industry fully occupied in state-controlled production to service a disastrous war, American industrial production was depicted as a creative vortex into which global raw materials were drawn and then extruded, not as state-owned vehicles of destruction but as private vehicles of pleasure. In these terms the article flagged the emergence of a new material framework for both personal and national identities and positioned the automobile as a vehicle for knowing what it meant to be a (privileged white) American at that particular moment in history. By borrowing its imagery from English expressions of pride in an empire over which "the sun never set," the editorial foreshadowed emerging forms of twentieth-century colonial relations that would come to replace the old empires of Europe. It was a worldview that confirmed a belief in the superiority of Western cultures in general and an emergent United States in particular. Such narratives helped to shape the attitudes toward and interactions with the peoples that wealthy motorists encountered as they circulated across the globe in their "cosmopolitan mechanisms."

It was only a few years after the turn of the century that numbers of men and women set out to "conquer the globe" in automobiles. Boston couple Charles Glidden and his wife, whose wealth derived from the Bell telephone company, became one of the best known of the early "world tour" motorists. They visited most continents between 1901 and 1908 in their Napier touring cars, visiting twelve thousand cities and traveling almost fifty thousand miles—much of it with wheels specially converted to travel along railway tracks. Less leisurely enterprises soon followed, with a Paris-to-Peking automobile race organized in 1907 and a New York–to–Paris race in the following year. Most notable among early female global tourists was Mrs. Harriet Fisher, a widowed businesswoman from Trenton, New Jersey, who in 1909 traveled in style in her luxurious Locomobile. She was aided by her chauffeur, an English manservant, and an Italian maid, and they "girdled the globe at leisure" in a twenty thousand–mile tour that took a little over a year.

Around the world driving tours attracted publicity for some decades, and by the 1920s numerous automobile parties were to be found hauling their cars though rivers, over mountain peaks, and across desert sands and inching along bridges that had not been designed with such traffic in mind. From a British perspective five "classic" long-distance routes had been identified by the 1920s—around Australia, England to Australia, Cape Town to Cairo, Algeria to Cairo, and the Sahara crossing. According to one motoring historian, the years between 1922 and 1928 saw those five routes "swarming" with pioneering motoring expeditions, in which adventurers not only pursued personal challenge and pleasure but also sought to bring credit to their national motor industries and open out colonial territories to modern transportation.[3]

The journeys were eminently susceptible to measurement of all kinds—the first party to "conquer" a particular route, miles traveled, the length in months or years on the track, numbers of punctures or tires ruined, gallons of fuel and oil consumed, axles broken and leaf springs repaired, rivers forded, quantities of trees cut down to construct a road, bouts of fever endured, and weeks that passed without seeing another white face. Stories about such carefully calibrated journeys provided a twentieth-century register for what Michael Adas has termed "machines as the measure of men." In the context of Western colonial expansion, Adas argues, Western people's technological and scientific achievements served them with unambiguous proof of the superiority of their cultures and constituted the key to understanding colonial travel narratives.[4] Even though in the most extreme of those journeys it would be closer to the truth to acknowledge that early global motorists had their automobiles hauled over mountains, deserts, and rivers by the muscle power of local people and their animal teams, as much as by the power of modern technology, it was Westerners on the move who were cast as the stories' heroes.

The 1930s was the high point of British colonialism in sub-Saharan Africa, as Britain consolidated its interests during the interwar period. In the 1800s British interest in the "Dark Continent" had been largely confined to the Cape colony to the south and Egypt to the north, the two separated by five thousand miles of territory largely unexplored by Europeans. By the late nineteenth century, however, during a period of crisis in British industrialization sometimes known as the "second industrial revolution," Africa began to gain prominence over Asia as a colonial possession. By then the advantage that Britain had accrued from its earlier industrial revolution was coming to an end. With Germany and France as well as the United States having undergone rapid industrialization, Britain was losing its leading position in the industrial world, and expanding trade and commerce in Africa was seen as integral to renewed British growth.

"Cape-to-Cairo" became shorthand during the late nineteenth century for a proposed rail route that climbed the mountains behind Cape Town, passed along the resource-rich and fertile high tablelands to the strategically crucial chain of great lakes, then through the swamps of southern Sudan and along the Nile Valley into northern Egypt. It was imagined as a transport "backbone" that could tie together British interests from north to south, confirming British supremacy over its European colonial rivals and bringing southern Africa into the Mediterranean orbit. In the cartographical convention of an "all-red," British line of transportation stretching the length of the African continent, the Cape-to-Cairo route would speak to the world of the persistence of British engineering superiority and colonial backbone into the twentieth century.

Even as British interests expanded in Africa, however, European colonialism was becoming a much less secure project. With the rise of a colonial middle-class and nationalist sentiment in the interwar period, the myths of British racial, cultural, and industrial superiority were not as self-evident as they had seemed at the turn of the century. Those social and economic changes inflected the ways in which stories about white women's automobile trips through Africa were told, ensuring that they were as concerned with alterations in the world order as they were about individual aspiration or intrepid lady adventurers. Women's travel narratives were not just about changes in ideas of gender; they were also thoroughly steeped in contemporary debates about technology, nation, and empire, in which the persistence of postwar problems in the British economy, nationalist aspirations within colonial territories, and the increasing industrial domination of the United States in global markets were all being played out.

The British imperial dream of a continuous Cape-to-Cairo rail line was never realized. The much less infrastructure-hungry solution of motoring was taken up in the years following World War I, though it remained a tremendously arduous undertaking and tackled by only the most adventurous travelers until after World War II. One of the earliest attempts to drive a car along the route was recorded in 1914, when a party led by a British army officer, Major Kelsey, took nine months to move an ornate Scottish Argyll car a little over one-fifth of the way, two thousand miles from Cape Town to Broken Hill (Kabwe). Kelsey's attempt came to an end when a leopard killed him.

It was another ten years before the first successful British trip, when Major Court Treatt's party made it through after sixteen months of the most astounding hardship, driving, pushing, and dragging two British Crossley trucks the length of the continent, never deviating from British territory. Stella Court Treatt, one of the "bright young things" of fashionable London society, produced a popular book about the trip, *Cape-to-Cairo: The Record of a Historic Motor Journey* (1927).[5] She recorded their optimistic departure in grand colonial style, wearing *sola topees* and tailored tropical suits. They carried Jaeger blankets, a gramophone with jazz records for dancing as well as music of modern Russian composers, and folding camp chairs inscribed with their names—most of which they were forced to discard along the way. Even Stella Court Treatt's loyal and admiring account of her husband's leadership could not disguise the horrors of the trip. The party spent four months covering the less than four hundred miles between Bulawayo and Victoria Falls through heavy rains, wallowing in deep mud, and totally dependent upon the help of the local population to keep them moving. They were hungry and on short rations, permanently wet, frequently ill, plagued by scorpions and ants, their clothes and bedding mildewed. Their film of the journey was released in London cinemas in March 1926, and, though the footage has now been lost, a reviewer described its content. It showed hundreds of Africans helping Major Court Treatt's expedition, hauling ropes to drag his two cars over mud, through rivers and along the edge of precipices. One tribe was forced to drag the wood necessary to build a perilous bridge for over ten miles. And "sometimes there was neither bridge nor raft, and the cars were dragged under water along the riverbed."[6]

When the Court Treatt party arrived in London, their journey was celebrated as a triumph of the British spirit and the superiority of British engineering, but more sober assessment has considered their determination to drive antiquated and wholly unsuitable British vehicles that were grossly overloaded, every yard of the route on "British soil," to be a folly of British imperial adventurism. But stories about it, including the film the group made of the journey, served to enhance the mystique of the Cape-to-Cairo route and made the hardships of later trips pale into insignificance. No subsequent party could possibly display such British pluck, and nobody could claim to have endured such an ordeal as the Court Treatts had.

Court Treatt's party returned to London three months before the General Strike of May 1926, and that intense period of class conflict provided the heated environment in which Stella Court Treatt's book was produced. Motorcars were welcomed as a weapon against working-class militancy during the General Strike, and a government agency coordinated motor vehicle owners and drivers as volunteer strikebreakers to run supplies and to provide public transport. Cars proved to be an effective and flexible response to working-class organization, especially to the power of the railway unions, and Stella Court Treatt's account of one of the major dramas of their expedition, their struggle with "labor relations" in Africa, resonated with the domestic struggles being fought out in British industrial relations as she wrote. She detailed their difficulty in directing the uncooperative native labor that had been assigned to them by British administrators as a micropolitics of imposing their will on a resisting colonial proletariat. The reluctance of tribesmen to build roads through swampland for the traveling party resonated with the domestic struggles to keep miners at work in British coal shafts.[7]

In admiring descriptions that harked back to Victorian and Edwardian accounts of imperial travel, she wrote that Major Court Treatt resolved their difficulties by physically imposing colonial power over the "Red Dinkas," reducing them to feminized men or the status of children. Order and British prestige were reestablished in the restive Sudan by brute force, just as decisive action resolved working-class insubordination at home. But, as Elizabeth Collingham has argued in the context of the British raj in India, such bald physical attempts to impose notions of race superiority through aristocratic assumptions of rule contributed to the downfall of British colonial power.[8] The arrogant assertion of white masculine power was an ineffectual response to the complexities of colonial race relations and rising anticolonial movements in the twentieth century. By the 1920s even the British themselves were losing confidence in the possibility of physically imposing that British sense of superiority.

The physical imposition of authority had rarely been the province of women, and by the twentieth century those who adopted that stance found little admiration in its telling. One British woman who tried to do so was writer and adventurer Diana Strickland, who drove seven thousand miles across northern Africa, from Dakar in the West to Maswa in the East, in her Wolverhampton-built Star motorcar in 1927–28. Strickland, who was characterized in the press as "a modern woman to the fingertips," a "wealthy society beauty," "tropical explorer," and "big game hunter," drove across the continent on a thirteen-month journey that was dogged by endless misfortune. Her navigator died within three days of contracting a severe form of malaria known as blackwater fever, her mechanic returned to England ill, and French authorities did all they could to prevent her from crossing the Sahara alone. Unlike the earlier walking trek she had organized through the Congo, Strickland never published a narrative of the journey, though she sent regular dispatches back to the company that had manufactured her car. The Star Engineering Company fed the Wolverhampton press glowing reports of her progress across Africa, but her journey failed to cohere as a story or to strike a sympathetic chord. Strickland's accounts of her heroism, in which she told of using her car to push down a native hut after villagers refused to give her water, were modeled on the white "memsahib" style of colonial travel and struck a jarring, outdated note. Her narrative was too close to the stories of imperial travel in the Victorian and Edwardian eras, and the press was inclined to lampoon her pretensions rather than reproduce a ripping yarn of colonial adventuring.

By the end of the 1920s in Britain the turmoil of the General Strike had faded into insignificance against the disaster of the Great Depression. In that new industrial crisis, in which middle-class people were also affected, it was difficult to present working-class recalcitrance as the major issue preventing British recovery. Those altered conditions in the representation of colonial race relations, the severe problems in the British industrial landscape, as well as increasing nationalist pressures from within the colonies themselves were reflected in Margaret Belcher's quite different narrative of her six-month journey from Cape Town to Cowley. Her account suggested a coming version of British colonialism that was cognizant of the empire's dwindling global status and responsive to the changing conditions of postwar Africa.

Margaret Belcher and Ellen Budgell were accompanied by their female friend J.O.M. Day, who traveled with them as far as Nairobi, acting as cook to their driver and mechanic. Although separated by only a few years, Belcher's travel narrative, published as *Cape to Cowley via Cairo in a Light Car* (1932), was quite different from that produced by either Stella Court Treatt or Diana Strickland. While remaining immersed in the assumptions of a colonial order that maintained "Englishness" as a superior category, Margaret Belcher's account was altogether more modest, self-reflexive, and conscious of the ironies of colonial adventuring. Given the impact of the Depression, their much more marginal status within the colonial enterprise, and the additional pressures they faced traveling without white men, Belcher documented a less confident and less strident version of traveling Englishness, recording a "battlers" nationalism in which British imperialism was under pressure, less certain of its global standing and industrial future.

Their 1924 model Morris Oxford touring car had cost only sixty pounds—the price of a return sea passage from Cape Town to Southampton—and had already been driven twenty-five thousand miles before they bought it. Its small size and worn appearance, the car overloaded with items of luggage and camping equipment and with "doors that refused to shut without the aid of a rope" (52), did not easily lend itself to a heroic narrative of imperial adventuring, as Stella Court Treatt had so carefully presented. Instead, the women invested their actions with quite different meanings, manifested in the body of their car itself. They named it "Bohunkus," which they translated as "a tramp," the only male in the party. Imagined on the scale of the human body, rather than as a piece of conquering technology, Belcher wrote that they ceased to regard it as merely a car but considered it an "old and trusted friend." They had BOHUNKUS: CAPE-TO-CAIRO painted on the door, and Belcher even "affectionately dedicated" her book to "his" memory.

It seems that the name came to them from African-American music, then beginning to circulate around the world through radio and the recording industry. The great New Orleans jazz clarinetist Johnny Dodds had recorded "Bohunkus Blues" in 1926, but better known was the 1928 Columbia recording of "Bohunkus and Josephus" by the Birmingham Jubilee Singers, an a cappella gospel quartet from Alabama, whose distinctive style was a product of the migration of rural African Americans to the industrial mill and mine settlements of Jefferson County in the early twentieth century. "Bohunkus and Josephus" was a wry but buoyant song about

End of a long journey. Margaret Belcher and Ellen Budgell welcomed to the Morris factory at Cowley, where "Bohunkus" had been built. From T. C. Bridges and H. Hessell Tiltman, *The Romance of Motoring* (London: George G. Harrap and Co., 1933). By permission of the National Library of Australia.

two brothers and their shared poverty. Sung without instrumentation, it featured rhythmic, close harmonies in an arrangement in which the lead vocal was passed from singer to singer:

> And these two boys they bought a horse
> An' this ole horse was blinded
> Josephus rode on front of him
> Bohunkus on behind 'im.[9]

The women used the song as a framing metaphor for their female expedition, in which all they were able to assert was a kind of adventurism at the margins. Their journey was entirely devoid of the imprimatur of an official colonial expedition; it was a leaderless, democratic affair that was funded only by their modest incomes. They were unable to lay claim to the pioneering status of the Court Treatt party, they did not have any commercial backing, nor did they enjoy the cachet of the London social set in which the Court Treatts moved. Thus, although it was deliberately an "all-British" trip and following the "all-red" route across Africa, their journey was not dignified by the kind of civic ceremonial extended to earlier parties as they left Cape Town, and it was not honored by high-society receptions when they reached England. And, because the women were prepared to have the car shipped by rail or river steamer over sections where automobile travel proved too difficult, they also could not pretend it was a semi-military assault on hostile territory, as the Court Treatts had.

Quite the opposite—even though they were proud of being the first women to complete the journey unaccompanied by men, the two carefully disavowed pioneering heroics and emphasized the makeshift nature of their journey, in which their movement through the country was ungainly and comical as often as it was admirable. As Margaret Belcher pointed out, their departure on April Fools' Day preempted any grand claims, their actions merely a desire for pleasure and new experiences. "The call of the road was strong and not to be denied," Belcher declared, "and we set out purely in the spirit of adventure" (7). That avowed spontaneity and apparent lack of official purpose did not mean, of course, that they did not ascribe an overarching significance to their actions, for there was a broader historical context to their story, which they carefully employed to shape its meanings.

During the interwar years in British dominions such as South Africa and Australia, the economic act of buying a car was at the same time a statement of national allegiance. The market for British cars in South Africa, which had been strong before the outbreak of World War I, collapsed when imports from Britain ceased in 1914, and automobiles from the United States and Canada filled the vacuum. It was a trend that continued throughout the interwar period, so that by the outbreak of the next war North American cars constituted 80 percent of all imports into South Africa.[10] It was not simply that American cars were more readily available to South African consumers at a lower price than the British cars, but they were better suited to the harsh conditions of colonial motoring. Even though British firms attempted to recoup their prewar markets by producing special colonial or dominion models in the 1920s, they were only marginally different from those made for domestic consumption. British cars were not as robust or powerful, and they had less ground clearance and narrower track dimensions than American cars. On unpaved roads British cars could not easily follow the tire tracks made by previous travelers, and motorists were forced to drive for miles with one pair of wheels on the track and the other bumping along the higher ground between the tracks.

Shortcomings in the British product and appeals to "buy British" in advertising campaigns meant that the purchase of a British car could be read as an act of imperial loyalty during the interwar years, a vote of support for the industrial products of the "old country" against an increasingly ascendant American industry. The two women embraced that interpretation, imagining Bohunkus as a vehicle of British endurance, a brave and tattered reminder of British engineering superiority that once was and might be again. On their return to the factory in Cowley, where Belcher's family lived, the women gave interviews to local newspapers. Mindful of the importance of the Morris Motor Company in the Oxford economy and the threat local products faced from American mass production, they proclaimed that British engineering and British workmanship had not once let them down. It was a rather generous assessment of the reliability of the car, as Belcher's book revealed they had a great deal of trouble with the clutch, suspension, and rear axles. Her unpublished journal catalogued even more mechanical mishaps than the book subsequently admitted. Their first breakdown, a broken rear axle shaft, had occurred even before they had left Cape Town proper, only five miles into their journey. Bohunkus had to be towed back for repairs by a truck from the American Texaco Company, and Belcher joked that they had bribed the driver substantially to keep their unfortunate start a secret.

Beyond a declaration of confidence in a beleaguered British manufacturing industry, Bohunkus, as a less-than-prepossessing male in an all-female expedition, gestured toward changes that were taking place in the gendered order of imperial rule, in which middle-class white women were beginning to breach the masculine culture of the Colonial African Service and take a greater part in nongovernmental institutions. Although the expedition did not have governmental or commercial backing, the three women were nevertheless embedded within female networks that enabled them to devise their own ceremonies and differently formalize their location within a "British way of life" that stretched from England to the Cape and back again. And, while the colonial structures they used were decidedly peripheral in comparison to those available to male adventurers, they were no less rooted in a sense of community in which home and empire were linked in ways that gave their "wanderings" both a personal meaning and a broader political significance.

The women represented themselves as new expressions of white, colonial femininity, far removed from the earlier stereotype of settler wives who could "recline in a deck chair with a cold drink and leave everything to the natives" and quite different, again, from privileged society adventurers such as Diana Strickland and Stella Court Treatt (32). Margaret Belcher was a caseworker with the Cape Town Society for the Protection of Child Life and later its general secretary. The society, founded in 1908, was by the late 1920s a prominent welfare organization that enjoyed royal patronage, well-connected committee members, and numerous voluntary workers. It largely focused its activities on poor white children, managing residential institutions, adoptions, health care, and fresh air camps in the hope of preventing them from slipping beyond the bounds of white prestige. Ellen Budgell had been born in Croydon, Surrey, in 1892 and worked as a taxi driver in Cape Town. She had been a motor transport driver in the Women's Royal Naval Service in London during World War I and lived with Belcher in the same private hotel. In Belcher's book and in their newspaper interviews the women presented themselves as single women whose sense of identity revolved not around heterosexual domesticity but in their paid employment and public positioning as avid leaders in a vigorous Girl Guide movement, which by the late 1920s had thousands of members throughout Africa.

Guiding enabled the women to steep their journey in the ideology, rituals, and structures of an international movement that was linked to broader colonial structures. It provided them with a context within which they could develop their own expressions of female independence. They choreographed their own farewell ceremony as they left Cape Town, decorating Bohunkus with the colors of the Rondebosch Guide troop, for which Belcher was district captain, and arranged for uniformed Rangers, Guides, and Brownies to tow them the first yards of their journey through a lane of cheering spectators. The women carried the Rondebosch colors on their journey and displayed them at significant moments during the trip. They were "thrilled to the core" to be the first car permitted to use the new Victoria Falls Bridge and enacted a ritual in which they tied their troop colors across the center of the bridge, breaking them as they drove their car through. When they were forced by mechanical breakdown or flooding to make a more permanent camp, the women erected a flagpole and structured their day around the semi-military activities of a Guide outpost. The brass trefoil, a Guiding symbol they had fixed onto Bohunkus's radiator, enabled them to be identified throughout Africa, so they were drawn into a supportive network of Guide troops along their route.

Like Scouting, Guiding brought to British children a movement structured around fantasies of a colonial frontier, which founder Robert Baden-Powell had formulated out of his experience in the South African wars at the turn of the century. Its emphasis was on teaching "over-civilized" children a code of ethics and simple outdoor skills such as woodcraft, camping, and tracking, which avowedly originated in traditional African knowledge. The activities were considered an antidote to modern social ills. But the return of that "knowledge" to Africa in the form of a colonial Guiding movement accentuated the contradictions of romanticizing the "noble primitive" in a context in which a stated aim of the colonial project was to uplift and civilize the native out of that degraded state of primitivism and in which Guide troops were carefully segregated along racial and ethnic lines, despite the internationalist and multiracial principles of the movement. As Margaret Belcher's narrative revealed, Guiding in Africa, unlike in Britain, placed its adherents in situations in which fantasy collided with reality, so that the incongruities of imperial ideologies were apt to surface in ways that were not easy to ignore. As the three Guiders set off from Cape Town in a down-at-the-heel and unreliable Bohunkus on April Fools' Day, they were cast into situations in which the fragility of notions of white superiority was graphically played out.

It was in the incidental details of their daily encounters with African men that Belcher's narrative suggested the ways they were compelled to recognize some of the limitations to their knowledge and power as European women in a landscape that was patently not theirs. They were stories told against themselves, brief glimpses into exchanges in which some African men, but no African women, were endowed with a degree of subjectivity that Stella Court Treatt's determinedly assertive narrative could not allow. One of the most important measures of a Guide's competency was cooking over an open fire, and Belcher related their struggles with fire throughout the course of their journey as comic scenarios that worked to undercut some of the pretensions of white Guiders in the bush. She recounted how they had once stopped to camp under a large tree and were struggling vainly with a fire that refused to light, when a "native" emerged from the darkness and pushed them gently aside: "Gathering a few specially selected twigs, he quickly had a blaze, and at last satisfied that the fire was well and truly alight, he flung his blanket about him, grinned broadly and strolled off into the bush" (29).

The women's constantly thwarted desire for a close view of African wildlife provided another theme that similarly expressed their awkward positioning upon the African landscape. In contrast to familiar tales of heroic "big game hunters," so central to stories of European travel in Africa, Belcher frequently reported their lack of success in seeing elephants, lions, or hippopotamuses. They were unable to "shoot" them with their camera, no matter how much they tried, and usually encountered only fresh spoor to show they had once again narrowly missed the iconic African experience. When they found lion paw prints around their camp one morning, the women filled them with salt and published a photograph of ghostly white prints in their book (149). The tracks were evidence of their exposure to danger, thrilling in retrospect and adding authenticity to their adventure but documenting their vulnerability as they cowered in their tent in the night hoping the lions would go away. It was a position quite different from the conquering bravura displayed in the classic photograph of Major Court Treatt proudly holding up the lifeless head of a leopard or of Diana Strickland happy to adopt the mantle of the big game hunter.

The comic undercutting of assumptions of imperial mastery that surfaced throughout the women's narrative was only one element of Belcher's story. They were flashes of candor, or self-awareness, that existed side by side with dominant endorsements of the European traveler's superior orientation as amateur anthropologist, able to generalize with certainty on the "character of the native" or the possibilities of African modernization. African men forever remained "boys" in Belcher's account; hospitable, helpful, and courteous toward the travelers but premodern, dependent, corrupted by contact with European settlement, and unable to wear Western clothes appropriately. African women were given a minimal role and were positioned as part of the environment, exotic and interesting but far removed from the active mobility and self-possession of the traveling women. Mastery of machines, a measure of European women's independence and advancement, placed them a world away from African women, who were represented as fixed in one location, subject to their men, and abjectly grateful to be left with their cast-off gas cans.

The women's mode of travel, unlike railway journeys or sea passages, entailed fleeting face-to-face contact with numerous indigenous men and some indigenous women on their home territory. Their style of travel was different, again, from the overland Cape-to-Cairo treks made some twenty years earlier, when travelers were carried along the route in shaded litters by teams of porters. Instead, Bohunkus enabled the women to achieve a measure of independence from African labor, though doing so required a degree of strenuous effort on their part. The result was that their own hard work was integral to their modernity and central to their pleasure and pride in the progress they made. Their labor was placed in the foreground in Belcher's narrative, and sliding though a patch of mud on the edge of a precipitous drop, racing the car over a rocky incline, or groveling under Bohunkus to repair a shackle bolt in the mud was described with relish: "We bounded from bump to bump, missing boulders by inches and cannoning over deep ruts and holes. The gradient was always approaching one in three, and we careered madly down, sadly thinking of the push that lay ahead to get Bohunkus to the top of the next rise" (113–14).

When the women's own efforts were not sufficient or Bohunkus let them down, there was always a ready supply of local labor to get them out of a fix. In an orientation quite unlike Major Court Treatt's determination to impose his will on recalcitrant workers, Belcher imagined a reciprocity of mutual satisfaction between parties—African tribesmen with their *tickey* (small coin) in payment and the women pleased to be able to confer largess and continue on their journey. To their interactions with indigenous men, Belcher ascribed something of a shared focus on the intense physicality of travel—pushing their car through mud, dragging it through rivers, towing it into campsites, and repairing it with whatever materials were at hand. She depicted gangs of African men plying trucks long distances, sometimes under the direction of white gangers and occasionally independently of them. The African truckers, who were experienced bush mechanics, qualified as "nature's gentlemen," who would always find time to assist stranded Guiders. White gangers, even those who appeared disreputable, could be relied upon to uphold European prestige in front of Africans. Race solidarity prevailed to classify the women as honorary white men, and the travelers enjoyed the certainty that they would not be denigrated as women drivers by the white gangers, a courtesy that might not be extended to them had they met in England.

As with the Australian overlanding adventurers, it was not so much the landscape they passed through that constituted the spectacle in their story but the women themselves. A large part of Belcher's story was devoted to imagining how people they encountered admired their courage and power. Of one trucking crew who had stopped to help them repair yet another broken rear axle, she judged that the "boys" of the gang, who were afraid of the lions and hated being far from their kraals, were amazed to find three white women happily camping out in the bush. On being asked what they were doing, the women explained as best they could: "The last we saw of them was a gang of round-eyed boys being wafted away on the tail of the lorry while we set to work to prepare camp. Not for anything would these boys have slept alone in the bush: therefore they thought we were quite mad; and we felt they would have had a wonderful tale to tell on their return to the kraal" (26).

Ostensibly offered as a humorous comment on a childlike lack of nerve that prevented Africans from fully inhabiting their country, it allowed Belcher to draw attention to the women's modern femininity without appearing to boast. The knowledge that they had safely returned to England confirmed that they were not mad after all and allowed them to enjoy being a source of entertainment for an African village. After all, they had brought home an even better story. More important, that backhanded affirmation of female pluck so proudly read into African eyes was a rhetorical device that exploited myths of race in the service of challenging myths of gender. For, in employing an attributed indigenous vision to challenge truth claims about fragile femininity for readers at home, Belcher was drawing on dominant ideas of racial difference. She harnessed her desire to contest notions of sexual difference to discourses of British racial superiority, so that African men's apparent acknowledgment of the women's superior nerve, as they calmly camped in country where local people were said to be unwilling to remain, was bolstered by an imperial commonsense that declared that it was courage and immunity from superstition, rather than, for example, an ignorant disregard for the local specificities of place, that allowed British people to be at home anywhere in the world.

The picture was more complicated, however, as Belcher admitted that they were, indeed, sometimes scared by the country they were moving through. She wrote that they spent sleepless nights banging pots and empty gas cans to keep lions at bay and practiced scrambling up trees so they might avoid charging wildlife. Even more sobering, they became dangerously lost in desert country north of Khartoum in the Sudan, where government officials had strongly warned them not to enter, and found themselves in serious trouble before they found their way back to the Nile. The women had a lucky escape from death, everyone agreed, and they contritely put Bohunkus on the train as they had been advised to do in the first place. In the end the women were open to the kind of accommodation that Major Court Treatt had refused to countenance. His wife had loyally made light of the search parties that the administrations of both the Sudan and Egypt were forced to send out when their party became lost in desert country north of Wadi Halfa.

Belcher believed that their experiences had conferred on her the authority to speak plainly about the realities of the road conditions in Africa. In the appendix to her book, named "Tips to Tenderfoots," she wrote in conscious contradiction to what she called the "tall stories" told about African travel, advising travelers against driving all the way across continent. Among advice on mosquito nets, quinine, and iodine, she wrote that motorists should not be misled into thinking that it was possible to drive all the way across Africa. The "All-Red route" was a fine dream, but the whole of the Sudan, from Juba in the South until Wadi Halfa in the North, was impassable for cars (230).

In fact, the closer they came to England, the less Margaret Belcher's narrative placed them in control of their own progress. By the time they reached Egypt, where a limited form of independence had existed since 1922, the police had entirely taken charge of their itinerary, and they had to acknowledge they were elsewhere, in a country where they were no longer able to do as they pleased. They were whisked through the country at great speed, driving day and night under police escort, being lodged in police barracks, taking sightseeing tours that had been organized ahead of their arrival; they were also expected to speak at official receptions in towns they passed through. It was a degree of courtesy and ceremony that they found flattering but subtly coercive and tiresome. Even France had its terrors, and the hub-hub of Paris traffic, Belcher confessed, was a nerve-shattering experience after the quiet roads they had just left. They sat and "gibbered on the outskirts" of the city until one of the local Morris agents arrived to guide them through.

As the title of her book—*From Cape to Cowley via Cairo in a Light Car*—suggested, Belcher's narrative was of two British places, two homes brought into a new kind of proximity by their travel. One was on the colonial periphery and the other somewhere slightly off-center. The space in between provided the women with an opportunity to reinvent themselves away from the limitations of femininity that existed at either home, even while their movement through it was inflected by their minor positioning within colonial structures and imperial discourses. The modesty of their claims, their recognition that they were marginal to colonial enterprises, and their orientation of accommodation to the obstacles they faced allowed them to acknowledge some of the tensions contained in colonial discourses and something of the complexity of their positioning within them. They could not pretend to be autonomous, nor could they adopt with the same degree of conviction the masquerade of British prestige and race superiority that Diana Strickland or the Court Treatt party so readily assumed. Instead, they were forced to recognize their dependence upon the power, knowledge, courtesy, goodwill, and assistance of others. That contingency was a crucial difference in their travel narrative and worked to deflate retrospectively some of the heroic posturing of adventurers who positioned themselves less ambiguously. In fact, women's ability to make those journeys marked the demise in the pleasure men could take in them. One historian of colonial motoring lamented that much of the adventure and glamour of transcontinental journeys had been dissipated by 1930, when "it had become possible for three unaccompanied ladies to drive an aged and well-worn Morris from one end of Africa to the other without any great hardship or excitement."[11]

Rather than simply playing a spoiling role, however, Belcher's story allows us to return to those less-nuanced narratives of "excellent colonial adventures" and to read back into them some of the disavowed anxieties and disquiet that colonizers carried with them. As a story of women on the move in a colonial setting, framed by Bohunkus, that self-consciously modest cosmopolitan mechanism, Belcher's narrative offers a performative account of African travel, in which gender, race, class, nation, and empire all emerge as shifting social categories, captured at a moment when the global automobile industry was beginning to take a new direction.

Conclusion

In the first few decades of the twentieth century, when automobiles were still brand-new, enterprising female motorists were featured by the press as exemplary and admirable technological actors. Like men, women loved automobiles and invested liberatory hopes in them. They welcomed them as vehicles of pleasure, mobility, power, independence, and freedom. Becoming a motorist in those early years involved not just new intellectual knowledge about how the machines worked but also a myriad of physical adjustments, large and small, that meant both men and women were obliged to develop new bodily habits and skills. Early motoring required changes in sensory perception, reaction time, comportment; proficiency with specialized tools; dealing with dirt; reading the flow of traffic, adopting new styles of dress; developing courage, nerve, "knack," stamina, and coordination; and tuning into the unfamiliar sounds and rhythms of the "mechanism." For women it was those bodily adaptations that constituted their major challenge as they incorporated the new technology into their lives. Unlike men, however, aspiring female drivers had additional work to do if they were to become proficient motorists. They were also forced to grapple with elusive and fraught questions of how to participate in a technological world that had been gendered male—a world in which their female bodies were too readily deemed to be out of place.

The contradiction between being welcomed as consumers of automobiles but classed as incompetent technological actors paradoxically provided women motorists with an identity from which they could fight back. Their struggles against disparaging judgments represented a rejection of Victorian terms of sexual difference and a determination to shape what it meant to be a modern woman. In the early decades of the century female motorists effectively worked to alter the meanings of femininity through their engagement with automobiles in the hope that they might be able to move in social worlds on terms that were similar to those of men of their own class and race. Their creativity in producing new versions of femininity lay in their articulation, advocacy, and performance of bodily practices that were in marked contrast to those from the generation before them. They knew that those practices made all the difference in the world to what their bodies could do, what they might become, where they might be, and how they might go.

Motoring women everywhere contested the rigid division between men and women in the field of automobile technology, but their ambitions to expand their lives through the consumption of automobiles or to earn a living through their motoring skills were expressed and represented differently throughout each historical moment and national context. This book's focus on particular people and events in an international context shows how early women motorists took on a different character in Britain, the United States, and Australia as well as in a transnational, colonial context. Female motorists were aware of the actions of others around the world, but how they took to automobiles, what they said about their activities, and the ways others represented what they did reflected the local context and the resources available to redefine and expand understandings of gender difference. Motoring women were claiming a stake in public life as technologically savvy women, but just what that meant in each country was far from uniform.

British women, given the national crises of World War I and the country's declining industrial and imperial power, commonly expressed their ambitions in terms of female patriotism. In response to those national crises women's alacrity in taking on employment, fashions, and comportments that had once been the domain of men provided them with new sources of personal identity while simultaneously undermining notions of a sharp line dividing masculinity and femininity. In the United States, where there had been no such national crisis and the century promised most citizens a degree of prosperity not previously imagined, women motorists generally did not adopt a style of patriotic female masculinity but were much more likely to represent their hopes for better lives through the commercial imperatives of commodity consumption. They put themselves forward as exemplary and admirable consumers—a move encouraged and supported by automobile manufacturers and their agents. In Australia, a nation formed only at the beginning of the century, women came to public attention as skilled and admirable motorists by aligning their actions with the national project of producing a young white nation that was able to justify its claim of fully occupying a huge, old country. In each nation the ways women supported their claims to motoring competence changed markedly throughout the first three decades. At the turn of the century in both Britain and the United States women rested their claims on overt declarations of class and race superiority, legitimizing their place behind the wheel through their wealth and social prominence. Toward the end of the 1920s, however, women of much lesser privilege were employing more democratic resources to represent their motoring activities.

As the stories in this book show, early women motorists' success in presenting themselves as competent technological actors was not sustained into the 1930s. A new cohort of women and a more conservative social climate followed the generation of women who reached maturity in the years surrounding World War I. In each of the countries in this study the younger generation was less inclined to advocate gender equality in public arenas, emphasizing instead the values of domestic life and maternal femininity. Young women turned away from the radical forms of gender experimentation and political militancy that had characterized the previous generation, placing less value on women's participation in public life on the same terms as men. The financial crash at the end of the 1920s dampened the consumer optimism that had been crucial to women's motoring identities and restricted the small business opportunities they had turned to their advantage. Women's garages and driving schools failed; their chauffeur and hire-car businesses struggled to remain viable; manufacturers less often produced advertising images that depicted women as autonomous or independent motorists; and the few surviving women's columns in the motoring press published picnic recipes and golfing news, rather than the encouragement or mechanical advice common in the earlier years.

Yet it would be too simple to conclude that the gains women made in those years were lost or that the experimental radicalism of the years surrounding the war was thwarted as women were forced back to "home and duty" by an economic downturn and masculinist backlash. Certainly, women of the next generation were careful to distance themselves from feminism, but, if they did return to the pleasures of domesticity, their conception of what that meant had changed, and it was not an idea of the kind of home or marriage that their mothers had experienced. Home in the interwar years took on an entirely different aspect from Victorian or Edwardian times, with women of all classes anticipating changes in their marriages that would give them greater power over their lives. Once a radical prospect, it became commonplace for women to look forward to a companionate rather than authoritarian relationship with their husbands. Women brought to their marriages an expectation of sexual fulfillment and an intention to bear fewer children; they anticipated a large say in the economic decisions of the household, especially in terms of commodity consumption; and they enjoyed a healthy and active physicality outside of the home, less hampered by the clothing restrictions of the previous era. Even though the next generation of women disavowed the feminist movement, what it meant to be a woman had shifted, altered by those years of self-conscious gender experimentation and feminist activism. Those shifts led to a profound reconfiguration of relationships between men and women in the following decades.

Just as what it meant to be a woman had been transformed in those years, in part through women's engagements with new technologies, so too the meanings of automobile technology changed. The ground of the technology had shifted substantially, so that cars were no longer the same objects at the end of the 1920s that they had been at the beginning of the century. They had ceased to be unfamiliar and unreliable machines, they did not require the same degree of mechanical expertise on the part of their users, and they had become everyday objects, falling within the reach of an increasing number of households, though at different rates in each national context. As it was frequently declared, particularly in the United States, women's enthusiastic response to automobiles from the earliest years had played a large part in turning them into more sociable, or "user-friendly," objects. When manufacturers introduced advances in automobile design—from electric self-starters and automatic gearboxes to rear vision mirrors or windscreen wipers—they frequently announced them as a response to the special needs of female consumers. Women's particular needs, it was said, exerted pressure on engineers to modify the crude design of early automobiles, turning utilitarian, masculine machines into items of consumption that suited the mass of nonexpert users, who in reality included both men and women.

Although early women motorists were credited with playing a role in the mass diffusion of cars by prompting manufacturers to reconsider their design, the definition of women as especially needy technological actors took on much more overtly hostile forms during the decades that followed. In the era of mass automobile consumption—when car design had become stabilized, when the social meanings of automobiles had reached a degree of interpretive closure, and when ownership was broadening to include men of all classes—disparaging stories and jokes about incompetent woman drivers increasingly dominated public culture. Those disdainful judgments squeezed out earlier voices that had advocated the possibility of female competence or declared that women could be more sensitively tuned in to the technology than men. From the 1930s moves by middle- and working-class men to capture the field of automobiles went together with increasingly virulent attacks on female motorists. These attacks, often disguised as humor, culminated in the ubiquitous anti–woman driver jokes of the 1950s and

1960s, particularly in magazines and newspapers targeted to a male audience. There were endless variations of the "wife at the wheel" jokes, in which women were depicted as regularly backing up into fire hydrants, incapable of parking in the garage without running through the back wall, or using the choke lever as a hook to hang their handbags on. Young women drivers were characterized as unable to concentrate on the task at hand or give a clear hand signal, and mothers-in-laws were regularly portrayed as nagging backseat drivers.

Unlike the privileged women motorists of the first three decades, the broad mass of women motorists in the middle years of the century did not have the degree of leverage in public debates to enable them to counter the savagely disdainful judgments about their motoring skills. As a result, women's aspirations to be female mechanics, professional drivers, or garage proprietors almost entirely dropped from public view. Despite the fact that the number of women drivers increased tremendously as the automobile industry expanded, women were only rarely accorded an honorable place in the citizenship of the road. In that changed environment, where female technological competence had largely become unspeakable, women motorists were placed under renewed pressure as objects of curiosity, unsympathetic observation, and restraint. It meant that they had to do a great deal of extra work—over and above men of their class and race—to become adroit motorists. Far from merely reflecting sexual differences in relation to automobiles, the increasing dominance of stories about women's incompetence in those middle years of the century created a cramped space for aspiring women motorists. The stories actively worked to produce female technological incompetence—the very differences that they posited.

Shifting stories in the public media about women's engagements with automobile technology are not easy to interpret, however, as they do not provide a simple route to understanding what actual female motorists did during this period or how they understood their motoring competence. Changes in women's engagements with cars from the 1930s and beyond can only be clarified by careful empirical research. What we can be sure of is that, in spite of the decline in the visibility of female automotive competence and in the face of a culture that disparaged their technical abilities, some women—such as the auto mechanics at the Alice Anderson Motor Service—continued to practice the mechanical arts without public acknowledgment. Many more female drivers daily continued to handle their cars with confidence and pleasure, though their skills were hardly recognized, and sometimes they even encountered considerable hostility in public forums. In the United States, as Ruth Cowan notes, middle-class women became the key drivers for their households in the middle years of the century. Although they were not necessarily behind the wheel when it came to making social visits or taking pleasure trips, women increasingly became the suppliers of routine transport services for their households, taking over a role that had once been the domain of working-class men, such as the deliverymen who had previously brought products to their door or servants who had chauffeured family members around town. For the middle-class woman an automobile became "the vehicle through which she did much of her most significant work, and the locale where she could most often be found."[1] A similar phenomenon of "Mum's taxi" also emerged in Australia and Britain, though not until much later, during the last quarter of the twentieth century.

Although by the 1930s competent female motoring had become, in large part, a practice without a public discourse, women of much lesser privilege learned to drive in increasing numbers, and some quietly continued to work as motor mechanics. By 1939, when another war was imminent, female motorists in all three countries, some of them former transport drivers and mechanics in World War I, once again began to be heard in public debates. They established a variety of voluntary training units and lobbied their governments to institute military detachments that could employ the young women they were training. In Australia the transcontinental motorist Gladys Sandford, who had been an ambulance driver in Britain and Egypt during World War I, set up a women's ambulance corps of National Emergency Services in Sydney as soon as war seemed inevitable and pressured a reluctant federal government to commission such voluntary units into the Australian military services.

During World War II all three countries formally established women's military service, and, as it had been in the previous war, motor transport work became one of the popular choices for new recruits. This time greater numbers of women from much less privileged backgrounds were trained as military transport drivers and mechanics. Newspapers were once again filled with stories about women's surprising mechanical facility and their skill in driving even the heaviest trucks. Systematic historical work is still lacking in this area, but my interviews with Australian women who served as transport drivers in the 1940s reveal that they vividly recall the satisfaction they took in their new capabilities and expanded mobility. Their pleasure comes across in the stories they tell and their talk about the reunions they continue to hold. For over sixty years they have preserved their fond memories in photograph albums and saved artifacts from those days—army driving licenses, lecture notes, and safe driving certificates. Instead of conforming to the expectation at the time that they should remain at home until they married, the women with whom I spoke relished the opportunity to travel widely throughout Australia, learning to maintain and drive all kinds of vehicles—motorcycles, staff

Another war, a new cohort of female military drivers and mechanics. Australian Women's Army Service trainees, 1944. Courtesy of the Australian War Memorial, negative no. 065304.

cars, ambulances, and heavy trucks — while also mixing with men and women of all classes who they might never have otherwise met.

When the war ended, the public recognition of women's technological competence was withdrawn even more quickly than it had been after World War I, and stories of incompetent women drivers returned during the 1950s, dominating public forums with growing intensity. Almost all of the women I interviewed were unable to find paid work that utilized their driving and mechanical skills, though some of them remembered dearly wanting to do so. They expressed great sadness at having to give up on their professional motoring ambitions and the camaraderie it had brought them. Although many of these women regretted losing the skilled work they had performed during the war years, they continued to drive as private motorists during the 1950s and well into their senior years, some of them repairing their own cars.

Sixty years after the end of the war, those Australian women I interviewed placed their happy experiences in the military in stark contrast to the suffering of many men who had served, especially those who had spent years interned in Japanese prison camps. At the end of the war female drivers were often assigned to meet ships carrying returning servicemen, and the women I spoke to recalled their shock at seeing the wretched condition of the men who survived their internment. So powerful was the national response to the sight that even those women who had not actually met the troop ships reminisced about it as if they had been there. Many of them remembered feeling a great deal of unease, even guilt, because their war experiences had been so very different and expansive. They expressed concern that the returning men, who were greatly surprised at finding self-assured young women so actively engaged in military life, should not have their masculinity placed in question by continuing evidence of female competency.

That disparity of wartime experience between many ex-servicewomen and the men who had survived the Pacific war made it difficult for the former transport drivers I interviewed, a large proportion of them married to ex-servicemen, to insist publicly on the full extent of their knowledge, skill, and pleasure in the work they had performed during the war. Their enforced silence added to the conditions that allowed the memory of women's war work to fade so rapidly from the national story. Many former service drivers stated that even their own children were barely able to accommodate the idea of their mother's wartime capabilities, despite the fact that they had seen the photographs and heard the stories. Their wartime experiences have nevertheless remained central to their sense of identity, but the evidence shows in subtle ways. The women with whom I spoke have countered the public forgetting about their wartime roles and achievements by fostering private expression about that important time—through self-published memoirs and, reunions and ex-servicewomen's commemorations and by continuing to cherish, all these years later, the moments in their lives when people close to them acknowledged their proficiency.

Like their mother's generation, these young women "returned" to home and marriage in the immediate postwar years with new ideals of romantic love and the expectation of being able to have an egalitarian relationship with their husbands. Their confidence and skills, greatly expanded by their war work and the gratifying part they had played in public life in the face of a national emergency, seemed certain to secure for them new recognition as female citizens. My research suggests, however, that many of these women were not able to realize their expectations. The shifts they had made in their understandings of gender difference and their hopes for romantic love were not matched by the men of their generation, many of whom had been greatly affected by their war experiences and unable to communicate their ordeals to their families. Perhaps some of their disappointments were recognized by their daughters, the generation who went on to instigate the feminist movement of the 1960s and 1970s.

The so-called second-wave feminists, those of my generation, found themselves reinventing an energetic feminist politics on a variety of fronts. It was a feminism happy to claim the name but one that frequently had little sense of connection to earlier forms of female activism. For the most part we were ignorant of the history of women's determination to expand their lives, but, even when we were aware of earlier feminist campaigns, we were often inclined to reject their concerns as misguided or vaguely embarrassing. Be they movements for suffrage, temperance, sexual purity, dress reform, maternal welfare, consumer protection, or international peace, most of us had a limited understanding of the historical contexts within which those campaigns had occurred. It was not until decades later, when women's history became a serious area of study, that we developed a greater appreciation of earlier forms of female activism.

Give a Girl a Spanner was the slogan that those of us who wanted to challenge the masculine culture of technology devised in the 1970s to encourage young women to become technologically confident subjects. It was a slogan that Hilda Ward, Dorothy Levitt, Gabrielle Borthwick, Dorothée Pullinger, Alice Ramsay, Ingeborg Kindstedt, Alice Anderson, Gladys Sandford, and Margaret Belcher—indeed, all of the motorists discussed in this book—might have endorsed in their own time, but we had no knowledge of their automotive ambitions or the work they had done to establish a climate of female mechanical competence. We assumed that no technologically confident women had come before. Press articles mistakenly portrayed us as the first female motor mechanics, recording an optimistic moment, which we all took to be the beginning of a new chapter in women's relationships to technologies. Yet the urge to locate an originating moment for women's mechanical ambitions has in fact been repeated since women first took to automobiles. While appearing to inaugurate a history, claims about women's new technological capabilities inadvertently erase that history, the claims relying on the erroneous assumption that women's technological competence has no past.

As the studies in this book reveal, women's ambitions to produce themselves as technically proficient subjects has a rich past, and any attempt to construct a linear narrative of progress will fail to do justice to the complexities of its expression. While there have been progressive changes in women's engagements with automobiles that have helped to expand their lives, there is only limited value in telling stories that simply attempt to record such advances. By relying on a monolithic concept of "women," stories of progress flatten out the variability that characterizes the experiences of different kinds of women, in specific locations, and at particular historical moments, and by employing the term *automobile* as if this object remained immutable over time, such histories fail to do justice to the radical changes taking place in the technology and meaning of automobiles over the twentieth century. An approach that tries to understand how the very categories we use to tell our stories also have a history promises a more finely grained understanding of the processes by which automobiles and gendered difference have been created in relation to each other and offers new ways to think about how automobiles have come to hold such a central place in our lives.

Notes

INTRODUCTION

1. Vera Marie Teape, "The Road to Denver (1907)," *Palimpsest: Iowa Historical Department* 61, no. 1 (1980): 3–11; Kathryn Hulme, *How's the Road?* (San Francisco: privately published, 1928), 13.

2. "Will Ride in Autos: Negroes of Nashville Prefer Them to 'Jim Crow' Cars," *Cleveland Journal*, 30 September 1903, 1, and 13 January 1906, 1; Theodore Rosengarten, comp., *All God's Dangers: The Life of Nate Shaw* (1974; rpt., Chicago: University of Chicago Press, 2000), 249–56.

3. Iris Marion Young, "Throwing Like a Girl: A Phenomenology of Feminine Body Comportment, Motility, and Spaciality," in *Throwing Like a Girl and Other Essays in Feminist Philosophy and Social Theory*, ed. Iris Marion Young (Bloomington: Indiana University Press, 1990), 141–59. In philosopher Moira Gatens's terms, "gender is a material effect of the way in which power takes hold of the body rather than an ideological effect of the way power 'conditions' the mind." *Imaginary Bodies: Ethics, Power and Corporeality* (London: Routledge, 1996), 66.

4. Nina E. Lerman, Arwen Palmer Mohun, and Ruth Oldenziel, "The Shoulders We Stand On and the View from Here: Historiography and Directions for Research," *Technology and Culture* 38, no. 1 (1997): 9–30.

5. Marcel Mauss, "Techniques of the Body (1934)," in *Incorporations*, ed. Jonathan Crary and Sandford Kwinter (New York: Zone, 1992), 459; Moira Gatens, "Power, Bodies and Difference," in *Destablising Theory: Contemporary Feminist Debates*, ed. Michelle Barrett and Ann Phillips (Cambridge: Polity Press, 1992), 120–37.

6. Rachael Herzig, "The Matter of Race in Histories of American Technology," in *Technology and the African American Experience: Needs and Opportunities for Study*, ed. Bruce Sinclair (Cambridge, Mass.: MIT Press, 2004), 155–70.

CHAPTER 1: MOVEMENT IN A MINOR KEY

1. "The Automobile in Newport," *Town and Country*, 27 September 1902, 15.

2. The Hon. John Scott Montagu, *Cars and How to Drive Them* (London: Car Illustrated Ltd., 1908), 48.

3. Montgomery Rollins, "Women and Motor Cars," *Outlook*, 7 August 1909, 859–60.

4. Hilda Ward, *The Girl and the Motor* (Cincinnati: Gas Publishing Co., 1908), 99; see also Hilda Ward, "The Automobile in the Suburbs from a Woman's Point of View," *Suburban Life*, November 1907, 269–71. My thanks to John Dorsey of the Boston Public Library for research help.

5. "Why Women Are, or Are Not, Good Chauffeuses," *Outing: The Outdoor Magazine of Human Interest* 44 (April–September 1904): 156.

6. Pemberton, *Amateur Motorist*, 27.

7. Mrs. A. Sherman Hitchcock, "A Woman's Viewpoint of Motoring," *Motor*, April 1904, 19.

8. *Times* (London), 20 June 1906, 14; *Automobile* (U.S.), 23 January 1904, 115–17.

9. Dorothy Levitt, *The Woman and Her Car: A Chatty Little Handbook for All Women Who Motor or Who Want to Motor* (1909; rpt., London: Hugh Evelyn, 1970), 16.

10. Montgomery Rollins, "Women and Motor Cars," *Outlook*, 7 August 1909, 859–60.

11. Robert Sloss, "What a Woman Can Do with an Auto," *Outing*, April 1910, 62–69; Mrs. Andrew Cuneo, "Why Are Women Drivers Few?" *Auto Era*, May 1908, 6–8.

12. Artemis, "Women A-Wheel," *Australian Motorist*, 1 January 1915, 434–35.

13. Mrs A. Sherman Hitchcock, "For the Motoring Woman and about Her," *Motor Car*, July 1911, 8.

14. Lady Jeune, "Dress for Ladies," in *The Badminton Library: Motors and Motor Driving*, ed. Alfred C. Harmsworth (London: Longmans Green and Co., 1902), 66.

15. Ladies at Brooklands Web site, www.hartlana.co.uk/bsarchive/imo73.htm.

16. "Chauffeurs Draw the Color Line," *Motor World*, 1 July 1909, 555; "'Queering' the Negro Driver," *Motor World*, 24 February 1910, 560.

CHAPTER 2: A WAR PRODUCT

1. *London Daily Mail*, 29 December 1908, 3; *London Daily News*, 25 December 1908, 6 and 29; *Times* (London), 29 December 1908, 5; *New York Times*, 27 December 1908, 2, *Motor Car* (U.S.), March 1910, 32.

2. "A Woman Van Driver," *Englishwoman's Review of Social and Industrial Questions*, 15 April 1891, 131–32; "Now London is to Have Women Cabbies," *New York Sunday Journal*, 1 August 1897, 21.

3. *Autocar* (UK), 4 May 1907, 661.

4. Gladys de Havilland, *The Woman's Motor Manual: How to Obtain Employment in Government or Private Service as a Woman Driver: All about Motor Driving and the Management of Motor Vehicles* (London: Temple Press, 1918).

5. Sue M. Bowden, "Demand and Supply: Constraints in the Inter-War U.K. Car Industry: Did the Manufacturers Get It Right?" *Business History* 33, no. 2 (1991): 241–67; Perkin, *Age of the Automobile*, 110–13 and 135–37; Roy Church, *The Rise and De-*

cline of the British Motor Industry (Cambridge: Cambridge University Press, 1994), 20–22.

6. Nigel Nicolson, *Portrait of a Marriage* (New York: Atheneum, 1973), 99.
7. *Autocar*, 17 May 1919; *Graphic*, 6 November 1920.
8. Some advertising images of uniformed women include Miss Stewart Custombuilt: *Motor* (UK), 13 August 1919, 35; *Autocar* (UK), 20 March 1920, 95; *Australian Automobile Trade Journal*, November 1926, 11; *Motor* (U.S.), July 1919, 95; Burberry Motor Cycling Suit, *Graphic*, 17 July 1920; Rudge-Whitworth advertisement, *Motor*, 8 November 1920, 87; North British Clincher Motor Tyres, *Australian Motor Car*, December 1919.
9. Ray Strachey, "Women and Machines," *Common Cause*, 17 March 1916.
10. M.M., "Women as Motor-Drivers and Mechanics," 606.
11. Frances M. Hodgson, "Motor-Hiring," *Woman Engineer*, December 1920, 49.
12. *Ladies Pictorial*, 20 May 1920. My thanks to Barbara Burman for alerting me to this image.
13. *Westminster Gazette*, 11 November 1921, 1; classified advertisements in *Westminster Gazette*, throughout December 1921.
14. Radclyffe Hall, *The Well of Loneliness* (London: Doubleday, 1928), 147.
15. Laura Doan, *Fashioning Sapphism: The Origins of a Modern English Lesbian Culture* (New York: Columbia University Press, 2001), xiv.
16. *Woman Engineer*, 5 December 1925. For Ibbotson, see "Woman Owner/Driver," *Woman Engineer*, 10 March 1927, 6.
17. Ray Strachey, *Careers and Openings for Women* (London: Faber, 1935), 142–43.

CHAPTER 3: A CAR MADE BY ENGLISH LADIES FOR OTHERS OF THEIR SEX

1. *Light Car and Cyclecar* (UK), 5 February 1921, 232.
2. M.H., "Views of the Lady Owner-Driver on Olympia," *Motor* (UK), 19 November 1919.
3. "The Automobile World," *Queen: The Ladies' Newspaper and Court Chronicle*, 23 October 1920.
4. Edgar N. Duffield "Road Trials No. 38: The 10.9 Galloway," *Auto* (UK), n.d.; held in Arrol-Johnston clippings book, Dumfries and Galloway Museum.
5. Edwin Campbell, "The Olympia and White City Motor Show," *Graphic*, 30 October 1920.
6. *Lady's Pictorial*, 10 November 1917.
7. Galloway Engineering Co., *The Works and Its Environment: Engineering for Educated Ladies*, n.d., Imperial War Museum, Women at Work Collection, pt. 5, reel 72, MUN, 17.3/2.
8. "Gentlewomen as Engineers: A New Profession for Women," *Gentlewoman*, 17 November 1917, 902–3.
9. There was a forty-four-hour working week at Tongland with a starting wage of twenty shillings per week, five shillings below the Ministry of Munitions minimum wage at the

time. Six monthly increments of five shillings were tied to passing through each stage of the curriculum. The company claimed that it was possible for women to earn as much as three pounds per week once they were fully trained, but probably few ever approached that figure. Apart from examinations, women would also be graded for attendance, punctuality, conduct, cleanliness of machines and benches, and accuracy of work. Toward the end of 1919 piecework rates were introduced. Costs were about one pound per week for a hostel and an additional nine pence per day for a works canteen lunch.

10. *The Spanner: A Monthly Journal for the Control of the "Nuts,"* no. 2, February 1918, 24–26 (privately held). This was the combined Heathhall and Tongland works journal before the *Limit* was published. My thanks to Neale Lawson of Dumfries for sharing this document and his extensive research collection with me. "A Feminist Munition Shop," *Autocar*, 6 October 1917, 335.

11. "A Feminist Munition Factory," *Autocar* (UK), 10 November 1917, 454–56, 468; cloth badge held in IWM, MUN.17.3/3; plaque held at Motor Museum, Banton-on-Water.

12. Thomas Charles Pullinger, "President's Opening Address: The Institution of Automobile Engineers," *Proceedings of the Institution of Automobile Engineers: Session 1917–1918* 12 (1917): 432.

13. "The Girl Engineer," *Kirkcudbright Advertiser*, 9 November 1917; and unmarked newspaper clipping in Pullinger family papers. I wish to thank Dorothée Pullinger's family, Yvette le Couvey and Lewis Martin of Guernsey, for generously allowing me to study their private family papers.

14. Report of an interview with Miss Rowbotham, Galloway Engineering College, 31 December 1918, IWM MUN 17.3/11. My thanks also to the Girton College Archives Department. The Kirkcudbright Academy offered Tongland women, for a fee, mathematics and machine drawing courses two nights per week.

15. "A New Profession for Women," *Limit*, Xmas edition 1919, 8–9. My thanks to Jim Allen, National Trust for Scotland librarian at Hornel House, Kirkcudbright, who discovered copies of the *Limit* for me.

16. *Kirkcudbright Advertiser*, 1 August 1919. For a feature article on the Tongland works, see *Kirkcudbright Advertiser*, 13 June 1919.

17. *Kirkcudbright Advertiser*, 29 August 1919.

18. Women's Engineering Society papers, Institution of Electrical Engineers Archive, series NAEST 92.

19. "Women as Engineers: Union's Fear of Cheap Labour," *Times* (London), 8 December 1920, 9; "Women as Automobile Engineers," *Times* (London), 20 April 1920, 11.

20. *Times* (London), 23 February 1921, 7; *Vote* (UK), 2 June 1922; "Engineering Works Run by Women: Staff, Plant and Co-Partnership," *Westminster Gazette*, 4 May 1922. See pages of *Woman Engineer* throughout the 1920s for references to former Tongland workers.

21. Letter to auto historian Tim Aymes, 11 September 1982. My thanks also to Mr. Aymes for allowing me access to his manuscript "Scottish Motor Cars and Their Makers." *Motor* (UK), 9 June 1923; "A Prominent Lady Engineer," *Beardmore News*, June 1923; Dorothée Pullinger's typewritten memoir held in family private papers.

CHAPTER 4: TRANSCONTINENTAL TRAVEL

1. *Motor* (U.S.), January 1923, 28.
2. Lillian Faderman, "Lesbian Chic: Experimentation and Repression in the 1920s," in *The Gender and Consumer Culture Reader*, ed. Jennifer Scanlon (New York: New York University Press, 2000), 153–64; Laura L. Behling, *The Masculine Woman in America, 1890–1935* (Urbana: University of Illinois Press, 2001).
3. Frank Donovan, *Wheels for a Nation* (New York: Thomas Crowell, 1965).
4. Ruth Schwartz Cowan, *A Social History of American Technology* (New York: Oxford University Press, 1997), 224–48.
5. Carroll Pursell, *The Machine in America: A Social History of Technology* (Baltimore: Johns Hopkins University Press, 1995), 239.
6. "A Skilled Chauffeuse," *Automobile*, November 1900, 220.
7. *Automobile* (U.S.), 26 April 1906.
8. Mrs. A. Sherman Hitchcock, "Woman's Realm of Motordom," *Motor Car*, January 1910, 13–14.
9. Mrs. A. Sherman Hitchcock, "Woman's Realm of Motordom," *Motor Car*, May 1910, 10.
10. "Suffragette Now Runs a Taxicab," *New York Times*, 3 October 1913, 3. For Chicago taxi drivers, see *Motor Car*, November 1911, 6, and July 1912, 9.
11. *New York Motoring World*, 10 January 1915; *New York Times*, 2 January 1915, 8; "Girl Typist Seeks Job as Chauffeur," *New York Times*, 23 April 1916.
12. Burr C. Cook, "Women as Auto Mechanics," *Illustrated World*, June 1916, 503–4; "Teach Women Now at Auto School," *New York Times*, 22 March 1917; *Motor* (U.S.), July 1920, 56 and 94.
13. *Motor Age* (U.S.), 21 October 1915, 26–27; Mrs. A. Sherman Hitchcock, "Women's Realm of Motordom," *Motor Car*, June 1910, 40.
14. Bertha H. Smith, "A Lady Garage Man," *Sunset*, September 1914, 560–64; Sarah MacDougall, "Once a Manicure Girl—Now a Ford Specialist: Lady Weighs Ninety Pounds and Runs an Auto Repair Shop," *Ford Owner and Dealer*, May 1923, 62–64; "Two Successful Saleswomen Prove It Can Be Done," *Automobile Topics*, 27 March 1926, 624–25.
15. "She Repairs Automobiles," *Blacksmith and Wheelwright* (New York), October 1915, 833.
16. Amanda Preuss, *A Girl—A Record and an Oldsmobile* (Lansing, Mich.: Olds Motor Works, 1916).
17. *Not on a Broom*, MS, held in biographical files, Blanch Stuart Scott, National Air and Space Museum, Smithsonian Institution, Washington, D.C.
18. *San Francisco Chronicle*, 8 August 1909, 36.
19. *Hartford Press*, 29 October 1910, 1.
20. *Sacramento Union*, 6 August 1909, 2.
21. *Toledo Daily Blade*, 3 June 1910, 16:3.
22. *Desert Evening News* (Salt Lake City), 21 July 1909, 2:3.
23. *Cleveland Plain Dealer*, 27 May 1910, 9.

24. *Motor Field*, June 1910, 32.

25. "Women in a Reliability Run," *Motor World* (Chicago), 14 January 1909, 17 December 1908; *Horseless Age*, 20 December 1908; *New York Times*, 20 December 1908, S4; 11 January 1909, 1; 12 January 1909; 13 January 1909.

26. Mrs. A. Sherman Hitchcock, "For the Motoring Woman and about Her," *Motor-Car*, October 1912, 39.

27. *New York Times*, 16 December 1914, 13; *Motor World*, 12 December 1914.

28. *Woman's Journal*, 15 June 1912; 27 July 1912; "Disgrace to the Cause," *Santa Anna Daily Register*, n.d.; "Suffs Weep Over Sneers at Hiking Garments," Huntington Library, Susan B. Anthony Collection, Scrapbook 15, 33.

29. College Equal Suffrage League of Northern California, *Winning Equal Suffrage in California: Reports of the Committees of the College Equal Suffrage League of Northern California in the Campaign of 1911* (San Francisco, 1913), 61.

30. *New York Times*, 2 April 1916, 21; 6 April 1916, 20; 7 April 1916, 20; 1 October 1916, 20.

31. Margaret R. Burlingame, "The Motor Car Pays Its Debt to Women: How the Motor Vehicle Has Aided the Long Fight for Suffrage, Now Happily Ending," *Motor*, July 1919, 38–39 and 108.

32. *Fine Arts Journal* (Chicago), October 1912.

33. Madelaine G. Ritza in the *Michigan Manufacturer and Financial Record*, qtd. in "The Feminine Influence on the Car," *Literary Digest*, 18 March 1922, 64–65.

CHAPTER 5: CAMPAIGNS ON WHEELS

1. "Mud-Stained Car Carries Women Workers from San Francisco," *New York Tribune*, 27 November 1915. Between 1910 and 1914 Washington, California, Oregon, Arizona, Kansas, Alaska, Montana, and Nevada all became suffrage states. New York, New Jersey, Pennsylvania, and Massachusetts all had suffrage amendments defeated in 1915.

2. The Fifteenth Amendment passed in 1870 granted voting rights to black men, and the Nineteenth Amendment of 1920 enfranchised all women.

3. In 1915 the NAWSA had almost three million members, while the figures for the CU have been placed at between thirty-five thousand and fifty thousand at the peak of its membership.

4. *Bulletin* (San Francisco), 19 September 1915.

5. *Herald* (Washington, D.C.), 5 December 1915.

6. L. W. Peck, "Over the Lincoln Highway to the Coast," *Sunset: The Pacific Monthly*, April 1915, 772–86.

7. Huntington Library, Charles Erskine Scott Wood papers, WD box 277, SBF to CESW, 29 November 1915. My thanks to the staff of the Huntington Library in Pasadena for their support and research help.

8. *Emporia Gazette*, 20 October 1915, 1.

9. Mabel Vernon to Alice Paul, Des Moines, 29 October 1915, Library of Congress (LOC), National Woman's Party (NWP) papers, reel 20.

10. Mabel Vernon to Alice Paul, Des Moines, 30 October 1915, LOC, NWP papers, reel 20.
11. SBF to CESW, 29 September 1915, Huntington Library.
12. SBF to CESW, 6 October 1915, Huntington Library.
13. *Des Moines Register and Leader*, 30 October 1915; *Lincoln Daily Star*, 28 October 1915; *New York Telegraph*, 8 December 1915.
14. *Kansas City Post*, 21 October 1915.
15. *New York Tribune*, 27 November 1915.
16. *Nebraska State Journal*, 28 October 1915.
17. *Cheyenne State Leader*, 8 October 1915, 1; *Kansas City Post*, 20 October 1915; *Kansas City Post*, 21 October 1915; *Topeka Daily Capital*, 25 October 1915.
18. *Providence Sunday Journal*, 12 December 1915, sec. 3, 9 (my thanks to the reference librarians of the Providence Public Library); *Searchlight* held in the Rider Collection of Brown University (thanks to Mary-Jo Kline of the Christine Dunlap Farnham Archives at Brown University); LOC photograph reproduced in Margaret Finnegan, *Selling Suffrage: Consumer Culture and Votes for Women* (New York: Columbia University Press, 1999), 67.
19. *Kansas City Star*, 21 October 1915.
20. *Des Moines Register and Leader*, 30 October 1915.
21. Huntington Library, Charles Erskine Scott Wood papers, WD box 277, SBF to CESW, 6 November 1915.
22. *Lincoln Daily Star*, 28 October 1915.
23. *New York Tribune*, 27 November 1915.
24. *Boston Post*, 5 December 1915; *New Haven Register*, 12 December 1915.
25. *Kansas City Star*, 21 October 1915.
26. *Wilmington Every Evening*, 3 December 1915; 4 December 1915, 1.
27. Alice Sheppard, *Cartooning for Suffrage* (Albuquerque: University of New Mexico Press, 1994), 182 and 221.

CHAPTER 6: "THE WOMAN WHO DOES"

1. *Melbourne Herald*, 18 September 1926, 1.
2. Victorian Public Records Service, *Proceedings of Inquest*, VPRS 24 1034/1926; Supreme Court of Victoria, Probate Jurisdiction, Record of Will: Alice Elizabeth Foley Anderson, Probate no. 211/726.
3. *Melbourne Herald*, 30 September 1.
4. "A Woman's Venture," *Woman's World* (Australia), 1 February 1922, 13.
5. *Woman's World* (Melbourne), 1 February 1922, 13.
6. Sands and McDougal's *Directory of Victoria*, Melbourne, 1919.
7. *Australian Motorist*, 2 June 1919.
8. *Melbourne Herald*, 18 September 1926, 1.
9. *Woman's World*, 1 February 1922, 13; and "Women in Unconventional Callings," *Adam and Eve*, 1 June 1926, 20–21.
10. "A Clever Invention," *Australian Motorist*, September 1919, 20.

11. *Australian Automobile Trade Journal*, January 1920; Victorian Automobile Chamber of Commerce, Melbourne, Records of the "A" Grade Certificate Board of Examiners, 1926. My thanks to the staff at the Victorian Automobile Chamber of Commerce in Melbourne.

12. Letter from Marie Martin (Edie) to Mimi Colligan, 1983. Marie Martin's wages were only two pounds and one shilling for a six-day week, with no prospect of a pay rise until she turned twenty-one.

13. "The Woman Who Does," *Home*, 1 December 1920, 74.
14. *Australian Motorist*, 2 June 1919.
15. *Woman's World*, 1 February 1922, 13.
16. *Home*, 1 December 1920, 74.
17. Letter from Lucy Johnstone to Mimi Colligan, 1 April 1983.
18. Letter from Bib Stillwell to the author, 7 May 1996.

CHAPTER 7: DRIVING AUSTRALIAN MODERNITY

1. Michael Terry, *Across Unknown Australia* (London: Herbert Jenkins, 1925), 17.
2. Unidentified clipping, Sandford Papers, Mitchell Library, Sydney. MLMSS 4884.
3. *Motorist and Wheelman*, 20 April 1926, 39.
4. *Perth Daily News*, 7 April 1926.
5. *Sydney Mail*, 13 January 1926, 40.
6. *Brisbane Daily Mail*, 9 December 1925.
7. *Australian Motorist*, 1 February 1926, 345.
8. *Perth Daily News*, 7 April 1926.
9. *Sydney Mail*, 31 August 1927, 19.
10. *Northern Territory Times*, 17 May 1927.
11. *Fremantle Advertiser* 14 May 1926; *Radio Record* (New Zealand), 26 August 1927.
12. *Fremantle Advertiser* 4 December 1925.
13. Sandford, *Journal*, 2, Mitchell Library. MLMSS 4884.
14. Sandford, *Journal*, 9.
15. Marion Bell's diary and autograph book, private family papers.
16. *Car*, 5 February 1926, 27.
17. *Melbourne Herald*, 20 October 1928.
18. *Official Journal of the Royal Automobile Club of Victoria*, 15 October 1927, 7.
19. *Radio Record* (New Zealand), 26 August 1927.
20. *Fremantle Advertiser*, 4 December 1925, 1.
21. Sandford, *Journal*, 16.

CHAPTER 8: MACHINES AS THE MEASURE OF WOMEN

1. M. L. Belcher, *Cape to Cowley via Cairo in a Light Car* (London: Methuen, 1932).
2. Darwin S Hatch, "Where the Car Comes From: Atmosphere of Every Land in Make-up of Finished Vehicle," *Motor Age* (Chicago), 27 April 1916, 5.

3. T. R. Nicholson, *Five Roads to Danger: The Adventure of Transcontinental Motoring* (London: Cassell, 1960), xi–xiv.

4. Michael Adas, *Machines as the Measure of Men: Science, Technology, and Ideologies of Western Dominance* (Ithaca: Cornell University Press, 1989), 4.

5. Stella Court Treatt, *Cape to Cairo: The Record of a Historic Motor Journey* (London: George G. Harrap, 1927).

6. "'Cape to Cairo': The Court Treatt Film," *London Times*, 30 March 1926, 12D.

7. Treatt, *Cape to Cairo*, 194–95.

8. Elizabeth M. Collingham, *Imperial Bodies: The Physical Experience of the Raj, c. 1800–1947* (Cambridge: Polity, 2001).

9. Birmingham Jubilee Singers, *Complete Recorded Works Vol. 2 (1927–30)*, Document Records: DOCD-5346.

10. R. H. Johnston, *Early Motoring in South Africa* (Cape Town: C. Struik, 1975), 121.

11. Nicholson, *Five Roads to Danger*, 95.

CONCLUSION

1. Ruth Schwartz Cowan, *More Work for Mother: The Ironies of Household Technology from the Open Hearth to the Microwave* (London: Free Association Books, 1989), 85.

Essay on Sources

This transnational study has drawn on a broad range of documentary sources: automobile literature from the United States, Britain, Australia, and South Africa as well as studies on twentieth-century women's modernity, the history of technology, commodity consumption, and feminist theory. In this section, I list some of the key archival sources as well as books, magazine, newspaper, and journal articles that provided the foundation for this study. Among recent published sources Virginia Scharff's groundbreaking study of early women motorists in the United States, *Taking the Wheel: Women and the Coming of the Motor Age* (New York: Free Press, 1991), is the most important.

Since Scharff's work appeared, an increasing number of books include a consideration of women's engagements with automobiles. They include Graeme Davison, *Car Wars: How the Car Won Our Hearts and Conquered Our Cities* (Sydney: Allen and Unwin, 2003); Kathy Franz, *Tinkering: Consumers Reinvent the Early Automobile* (Philadelphia: University of Pennsylvania Press, 2005); Clay McShane, *Down the Asphalt Path: The Automobile and the American City* (New York: Columbia University Press, 1994); Daniel Miller, ed., *Car Cultures* (Oxford: Berg, 2001); Sean O'Connell, *The Car in British Society: Class, Gender and Motoring, 1896–1939* (Manchester: Manchester University Press, 1999); Martin Wachs and Margaret Crawford, eds., *The Car and the City: The Automobile, the Built Environment, and Daily Urban Life* (Ann Arbor: University of Michigan Press, 1992); Julie Wosk, *Women and the Machine: Representations from the Spinning Wheel to the Electronic Age* (Baltimore: Johns Hopkins University Press, 2001).

For some more popular books, see John Bullock, *Fast Women: The Drivers Who Changed the Face of Motor Racing* (London: Robson Books, 2002); S.C.H. Davis, *Atalanta: Women as Racing Drivers* (London: G. T. Foulis, 1955); Leonard Henslowe, *Woman and Her Car* (London: Gentlewoman and Henslowe Press, 1919); Curt McConnell, *A Reliable Car and a Woman Who Knows It: The First Coast-to-Coast Auto Trips by Women, 1899–1916* (Jefferson, N.C.: McFarland and Co.,

2000); Evelyn Mull, *Women in Sports Car Competition* (New York: Sports Car Press, 1958); Miranda Seymour, *The Bugatti Queen: In Search of a Motor-Racing Record* (London: Simon and Schuster, 2004). For two wonderful collection of images, see Friso Wiegersma, *La Belle chauffeuse*, trans. Jan Michael (Amsterdam: V.O.C.-Angel Books, 1981); and Gilles Néret and Hervé Poulain, *L'Art, la femme et l'automobile* (Paris: Editions EPA, 1989).

There have been many academic articles and book sections on women and automobiles published over the past few decades. Those I found most helpful are Laura L. Behling, "The Woman at the Wheel: Marketing Ideal Womanhood, 1915–1934," *Journal of American Culture* 20, no. 3 (1997): 13–30; Michael L. Berger, "Women Drivers! The Emergence of Folklore and Stereotypic Opinions Concerning Feminine Automotive Behavior," *Women's Studies International Forum* 9, no. 3 (1986): 257–63; Barbara Burman, "Racing Bodies: Dress and Pioneer Women in Aviators and Racing Drivers," *Women's History Review* 9, no. 2 (2000): 299–326; Georgine Clarsen, "The 'Dainty Female Toe' and the 'Brawny Male Arm': Conceptions of Bodies and Power in Automobile Technology," *Australian Feminist Studies* 15, no. 32 (2000): 153–63; Ruth Schwartz Cowan, *More Work for Mother: The Ironies of Household Technology from the Open Hearth to the Microwave* (London: Free Association Books, 1989); Laura Doan, "Primum Mobile: Women and Auto/Mobility in the Era of the Great War," *Women: A Cultural Review* 17, no. 1 (2006): 26–41; Catherine Gudis, *Buyways: Billboards, Automobiles, and the American Landscape* (New York: Routledge, 2004); Beth Kraig, "The Liberated Lady Driver," *Midwest Quarterly* 28 (Spring 1987): 378–401; Grace Lees-Maffei, "Men, Motors, Markets and Women," in Wollen and Kerr, *Autopia: Cars and Culture*, 363–70; Meaghan Morris, "Fear and the Family Sedan," in *The Politics of Everyday Fear*, ed. Brian Massumi (Minneapolis: University of Minnesota Press, 1993), 285–305; Sean O'Connell, "Gender and the Car in Inter-War Britain," in *Gender and Material Culture in Historical Perspective*, ed. Moira Donald and Linda Hurcombe (London: Macmillan, 2000), 175–91; Ruth Oldenziel, "Boys and Their Toys: The Fisher Body Craftsman's Guild, 1930–1968, and the Making of the Male Technical Domain," *Technology and Culture* 38, no. 1 (1997): 60–96; Charles L. Sandford, "'Woman's Place' in American Car Culture," in *The Automobile and American Culture*, ed. David L. Lewis and Laurence Goldstein (Ann Arbor: University of Michigan Press, 1983), 137–52; Virginia Scharff, "Gender, Electricity, and Automobility," in Wachs and Crawford, *Car and the City*, 75–85; Martin Wachs, "Men, Women, and Urban Travel: The Persistence of Separate Spheres," in Wachs and Crawford, *Car and the City*, 86–100.

Books published by women about their early motoring experiences, most of them now hard to locate, were a crucial resource for this study. Those I have been able to identify are Eliza Davis Aria, *Woman and the Motor Car: Being the Autobiography of an Automobilist* (London: Sydney Appleton, 1906); Betty and Nancy Debenham, *Motor Cycling for Women* (London: Pitman, 1928); Ethellyn Gardner, *Letters of the Motor Girl* (Boston: New England News Co., 1906), though the providence of this as a book is unclear; Gladys de Havilland, *The Woman's Motor Manual: How to Obtain Employment in Government or Private Service as a Woman Driver* (London: Temple Press Ltd. War Service Manuals, 1918); Baroness Campbell von Laurentz, *My Motor Milestones: How to Tour in a Car* (London: Herbert Jenkins and Brentano's, 1913); Dorothy Levitt, *The Woman and Her Car: A Chatty Little Handbook for All Women Who Motor or Who Want to Motor* (1909; rpt., London: Hugh Evelyn, 1970); Hilda Ward, *The Girl and the Motor* (Cincinnati: Gas Publishing Co., 1908).

Women took to transcontinental motoring early and published a great many books on their travel experiences. See Emma Augusta Ayer, *A Motor Flight through Algeria and Tunisia* (Chicago: McClurg, 1911); Mary Crehore Bedell, *Modern Gypsies: The Story of a Twelve Thousand Mile Camping Trip Encircling the United States* (New York: Brentano's, 1924); Margaret L. Belcher, *Cape to Cowley via Cairo in a Light Car* (London: Methuen, 1932); The Hon. Mrs. Victor A. Bruce, *9000 Miles in 8 Weeks: Being an Account of an Epic Journey by Motor-Car through Eleven Countries and Two Continents* (London: Heath Cranton Ltd., 1927); Winifred Hawkridge Dixon, *Westward Hoboes: Ups and Downs of Frontier Motoring* (New York: Scribner's, 1924); Muriel Dorney, *An Adventurous Honeymoon: The First Motor Honeymoon around Australia* (Brisbane: Read Press, 1927); Harriet White Fisher, *A Woman's World Tour in a Motor* (Philadelphia: J. B. Lippincott, 1911); Effie Price Gladding, *Across the Continent by the Lincoln Highway* (New York: Brentanos, 1915); Anne Bosworth Green, *Lighthearted Journey* (London: Century, 1930); Louise Closser Hale, *Motor Journeys* (Chicago: McClurg, 1912); William Holtz, ed., *Travels with Zenobia: Paris to Albania in a Model T Ford: A Journal by Rose Wilder Lane and Helen Dore Boylston* (1926) (Columbia: University of Missouri Press, 1983); Edith Wakeman Hughes, *Motoring in White: From Dakota to Cape Cod* (New York: Knickerbocker Press, 1917); Kathryn Hulme, *How's the Road?* (San Francisco: privately published, 1928); Mrs. Patrick Ness, *Ten Thousand Miles in Two Continents* (London: Methuen, 1929); Caroline Poole, *A Modern Prairie Schooner on the Transcontinental Trail: The Story of a Motor Trip* (San Francisco: privately published, 1919); Emily Post, *By Motor to the Golden Gate* (New York: Appleton and Co., 1916); Mary D. Post, *A Woman's Summer in a Motor Car* (New York: privately published, 1908); Alice Huyler Ramsey, *Veil, Duster and Tire Iron* (Covina,

Calif.: Grant Dahlstrom Castle Press, 1961); Caroline Rittenberg, *Motor West* (New York: Harold Vinal, 1926); Vera Marie Teape, "The Road to Denver (1907)," *Palimpsest: Iowa Historical Department*, 61, no. 1 (1980): 3–11; Jeanie Lippitt Weeden, *Rhode Island to California by Motor* (Santa Barbara, Calif.: Pacific Coast Publishing Co., 1916); Edith Wharton, *A Motor-Flight through France* (New York: Scribners, 1908); Maude Younger, "Alone across the Continent," *Sunset Magazine*, May 1924, 43, 106–9; and June 1924, 25–27, 60. For an excellent study of twentieth-century women's travel, see Sidonie Smith, *Moving Lives: Twentieth-Century Women's Travel Writing* (Minneapolis: University of Minnesota Press, 2001).

A number of motoring magazines published regular women's motoring columns in the first decades of the century. Some of the most significant were, in Britain, *Motorist and Traveller, Motor, Queen* and *Autocar;* in Australia, *Motor World Australia* and *Australian Motorist*. In the United States Mrs. A. Sherman Hitchcock was one of the most prolific, writing her column from at least 1904 until 1913, and her work appeared in a variety of journals, including *Automobile, Motor, Motor Car,* and the *New England Automobile Journal.* Hitchcock also published in women's magazines such as *American Homes and Gardens* and Suburban *Life*. There were other women regulars, such as the Paris-based writer and illustrator Blanche McManus, who published in the *Automobile, Harper's Bazaar,* and *Motor*. A great many occasional motoring columns published by women, such as Gladys Beattie Crozier, "Practical Motoring for Ladies," *Ladies' Realm*, September 1905, 571–78, can be found scattered throughout the women's press, in feminist journals, and in sporting magazines.

There is a growing literature that applies new cultural approaches to automobility, often in a transnational context. While much of it does not foreground gender, this work offers productive new analyses: William E. Connolly, "Speed, Concentric Cultures, and Cosmopolitanism," *Political Theory* 28, no. 5 (2000): 596–618; Sara Danius, "The Aesthetics of the Windshield: Proust and the Modernist Rhetoric of Speed," *Modernism/Modernity* 8, no. 1 (2001): 99–126; Tim Dant and Peter J. Martin, "By Car: Carrying Modern Society," in *Ordinary Consumption*, ed. Jukka Gronow and Alan Warde (London: Routledge, 2001), 143–57; Edward Dimendberg, "The Will to Motorization: Cinema, Highways, and Modernity," *October* 73 (Summer 1995): 91–137; Mike Featherstone, Nigel Thrift, and John Urry, eds., *Automobilities* (London: Sage. 2005); Mikael Hard and Andreas Knie, "The Grammar of Technology: German and French Diesel Engineering," *Technology and Culture* 40, no. 1 (1999): 26–46; Rudy Koshar, "On the History of the Automobile in Everyday Life," *Contemporary European History* 10, no. 1 (2001): 143–54; Gijs Mom, *The Electric Vehicle: Technology and Expectations in the Automo-*

bile Age (Baltimore: Johns Hopkins University Press, 2004); Kurt Moser, "World War One and the Creation of Desire for Automobiles in Germany," in *Getting and Spending: European and American Consumer Societies in the Twentieth Century*, ed. Susan Strasser, Charles McGovern, and Matthias Judt (Washington, D.C.: Cambridge University Press, 1998), 195–222; Tom O'Dell, *Culture Unbound: Americanization and Everyday Life in Sweden* (Lund: Nordic Academic Press, 1997); Kristin Ross, *Fast Cars, Clean Bodies: Decolonization and the Reordering of French Culture* (Cambridge, Mass.: MIT Press, 1995); Wolfgang Sachs, *For Love of the Automobile: Looking Back into the History of Our Desires*, trans. Don Reneau (Berkeley: University of California Press, 1992); Jeffrey T. Schnapp, "Crash (Speed as Engine of Individuation)," *Modernism/Modernity* 6, no. 1 (1999): 1–49; David Thoms, Len Holden, and Tim Claydon, eds., *The Motor Car and Popular Culture in the Twentieth Century* (Aldershot: Ashgate, 1998).

For a selection of early books published to induct the "amateur motorist" into the arts of technological modernity, see E. T. Bubier, *How to Build Automobiles* (Lynn, Mass.: Bubier Publishing Co., 1904); C. W. Brown, *The ABC of Motoring* (New York: Wycil and Co., 1903); Alfred C. Harmsworth, *The Badminton Library: Motors and Motor Driving* (London: Longmans, Green and Co., 1902); John Henry Knight, *Notes on Motor Carriages: With Hints for Purchasers and Users*, Motor Reprints Series (1896; rpt., London: Orbach and Chambers, 1970); Victor Lougheed, *Motor-Car Handbook* (New York: Motor, National Magazine of Motoring, 1905); Maurice Maeterlinck, "On a Motor-Car," *The Double Garden* (New York: George Allen, 1904), 139–52; Maj. C. G. Matson, *The Modest Man's Motor* (London: Lawrence and Bullen, 1903); Richard James Mecredy, *The Motor Book* (1903; rpt., London: Hugh Evelyn Ltd., 1970); The Hon. John Scott Montagu, *Cars and How to Drive Them* (London: Car Illustrated Ltd., 1908); Max Pemberton, *The Amateur Motorist* (London: Hutchinson and Co., 1907); H. J. Spooner, *Motors and Motoring* (London: Jack's Scientific Series, 1904); A. B. Filson Young, *The Complete Motorist* (1904; rpt., East Ardsley, Yorkshire: E. P. Publishing, 1973); A. B. Filson Young, *The Happy Motorist: An Introduction to the Use and Enjoyment of the Motor Car* (London: E. Grant Richards, 1906).

The following are additional, chapter-specific sources.

CHAPTER 1: Early motorists: "Now the Motor Woman," *Horseless Age*, June 1897, 4; "Chicago Warns Women Automobilists," *Horseless Age*, 12 September 1900, 24; Mrs. F. P. Avery, "Touring in Horseless Carriages—A Few Suggestions," *Automobile*, January 1901, 5; "A Lady's Tour of 1,500 Miles," *Autocar*, 11 May 1901, 446; "An Expert Woman Driver," *Car Illustrated*, 18 June 1902, 123; J. L. Clark, "London to Liverpool and Back: A Lady Driver's Journey on a De Dion," *Autocar* (UK), 8 April

1905, 490; Hiram Percy Maxim, "Learning to Drive a Motor Carriage," *Horseless Age*, April 1898, 5–6; Milton Lehman, "The First Woman Driver," *Life*, 8 September 1952, 83–92.

CHAPTER 2: Early women's motor garages: for Alice Neville, see *Motor* (UK), 2 September 1913, 139. My thanks to Hannah Crowdy of the Worthing Museum and Art Gallery for research help on Neville. For Borthwick, see "Chauffeuse-Companion: Increasing Demand for Women Drivers," *Times* (London), 11 December 1915, 11; M.M., "Women as Motor-Drivers and Mechanics," *Common Cause*, 23 February 1917, 606. There were numerous advertisements for women's garages in *Common Cause*, *Gentlewoman*, *Woman Engineer*, and *Women's Industrial News*. See records of Borthwick Garages Ltd. PRO (Kew), BT 31/24082 and BT 31/24410/172658; C. Griff, "Engineering as a Profession for Women," *Common Cause*, 1 October 1915, 310–11; B.P., "A Women's Motor Garage," *Common Cause*, 4 August 1916.

For some contemporary accounts of women's wartime motor work, see Edith Bagnold, *The Happy Foreigner* (1920; rpt., London: Virago, 1987); Pat Beauchamp, *Fanny Went to War* (London: Routledge, 1940); Molly Coleclough, *The Women's Legion, 1916–1920* (London: Spearman, 1940); Olive King, *One Woman at War: Letters of Olive King, 1915–1920* (Melbourne: Melbourne University Press, 1986); Katherine Hodges North, "Diary: A Driver at the Front," in *Lines of Fire: Women Writer of World War I*, ed. Margaret R. Higonnet (New York: Plume, 1999), 188–96; Helen Zena Smith, *Not So Quiet . . . Stepdaughters of War* (1930; rpt., London: Virago, 1988).

For a discussion of female masculinities, see Laura Doan, *Fashioning Sapphism: The Origins of a Modern English Lesbian Culture* (New York: Columbia University Press, 2001). For X Garage, see Kate Summerscale, *The Queen of Whale Cay* (London: Fourth Estate, 1997), 56–60.

CHAPTER 3: Records of the Galloway Engineering Co. and Galloway Motors Ltd. are held in the National Archives of Scotland, BT2/9604 and BT2/11500; see also the Women at Work Collection, Imperial War Museum, London, Galloway Engineering Co. Ltd. Tonglands, pt. 5 reel 72, MUN, 17.3..

For contemporary articles on the Tongland factory, see *Engineering*, 9 November 1917; *Lady*, 8 November 1917, 40; *Vote*, 16 January 1920, 466; Galloway Engineering Co., *A New Profession for Educated Women: Engineering: Expert Opinions of a Notable Achievement*, pamphlet held in the Ewart Library, Dumfries, n.d.; *Kirkcudbright Advertiser*, 29 December 1916, 9 November 1917, 13 June 1919, and 16 December 1921; *Dumfries and Galloway Standard*, 28 September 1918.

The Kirkcudbright Town Council Minute Books for 1919 are held at the Stewartry Museum, Kirkcudbright. My thanks to David Devereux, curator of the museum, who helped to locate town records and private photographs.

Recent assessments of British women's wartime work are provided in Janet S. K. Watson, *Fighting Different Wars: Experience, Memory and the First World War in Britain* (Cambridge: Cambridge University Press, 2004).

CHAPTER 4: For support during the research of this chapter, my thanks go to Sandy Biery, Yates and Gail Hafner, and Lauren Kata.

Recent work on the links between the U.S. suffrage movement and consumer culture, see Margaret Finnegan, *Selling Suffrage: Consumer Culture and Votes for Women* (New York: Columbia University Press, 1999); Susan A. Glenn, *Female Spectacle: The Theatrical Roots of Feminism* (Cambridge: Harvard University Press, 2000); Barbara Green, *Spectacular Confessions: Autobiography, Performative Activism, and the Sites of Suffrage, 1905–1938* (New York: St. Martin's Press, 1997); Linda J. Lumsden, *Rampant Women: Suffragists and the Right of Assembly* (Knoxville: University of Tennessee Press, 1997); Sarah J. Moore, "Making a Spectacle of Suffrage: The National Woman Suffrage Pageant," *Journal of American Culture* 20, no. 1 (1997): 89–104.

On Hollywood's serial queen genre, see Jennifer M. Bean, "Technologies of Early Stardom and the Extraordinary Body," *Camera Obscura* 48 16, no. 3 (2001): 9–57; Shelley Stamp, *Movie-Struck Girls: Women and Motion Picture Culture after the Nickelodeon* (Princeton: Princeton University Press, 2000).

CHAPTER 5: An excellent overview of franchise in the United States is provided in Alexander Keyssar, *The Right to Vote: The Contested History of Democracy in the United States* (New York: Basic Books, 2000). For African-American women and suffrage campaigns, see Rosalyn Terborg-Penn, *African American Women in the Struggle for the Vote, 1850–1920* (Bloomington: Indiana University Press, 1998).

For differences between the National American Woman Suffrage Association (NAWSA) and the Congressional Union for Woman Suffrage (CU), see Edith Mayo, foreword, in *Jailed for Freedom: American Women Win the Vote* by Doris Stevens, ed. Carol O'Hare (1920; rpt., Troutdale, Ore.: New Sage Press, 1995), 23; Nancy F. Cott, *The Grounding of Modern Feminism* (New Haven: Yale University Press), 72; Carrie Chapman Catt to Alice Paul, 12 April 1915, LOC NWP papers, group 2, box 1, folder 14; Ruth Barnes Moynihan, *Rebel for Rights: Abigail Scott Duniway* (New Haven: Yale University Press, 1983), 206ff; *New York Sun*, 4 August 1915; "Keen Rivalry Bobs Up in Suffrage Camps," *North American* (Philadelphia), 2 December 1915; "Rebukes for Militants," *New York Times*, 27 May 1915, 5.

There were hundreds of newspaper stories of the suffrage automobile campaign of 1916, which may be accessed by following the activists' itinerary, with approximate dates of major press reports, given as month/day, as follows: San Francisco 9/22; Reno 9/25; Salt Lake City 10/5; Cheyenne 10/8; Denver 10/12; Pueblo 10/16; Emporia 10/20; Topeka 10/25; Kansas City 10/21; Lincoln 10/22; Omaha 10/28; Des Moines 10/30; Chicago 11/5; Indianapolis 11/8; Dayton 11/9; Detroit 11/15; Toledo 11/12; Cleveland 11/1; Buffalo 11/16; Rochester 11/18; Providence 11/25; New York 11/27; Philadelphia 12/2; Wilmington 12/4; Baltimore 12/4; Washington, D.C. 12/5.

The Lincoln Highway: Drake Hokanson, *The Lincoln Highway: Main Street across America* (Iowa City: University of Iowa Press, 1988); Lincoln Highway Association, *The Complete Official Road Guide of the Lincoln Highway* (1916; rpt., Sacramento: Pleiades Press, 1995).

For the CU national conference of September 1915, see the entire edition of the *San Francisco Bulletin*, 19 September 1915.

On Sara Bard Field, see Amelia Fry, "Along the Suffrage Trail: From West to East for Freedom Now," *America West* 6, no. 1 (January 1969): 16–25; Amelia Fry, "Sara Bard Field: Poet and Suffragist (1882–1974)," in *Bancroft Library Regional Oral History Office: Suffragists Oral History Project* (Berkeley, Calif.: Regional Oral History Office, 1979). I am indebted to Amelia Fry for generously making her files available to me. Charles Erskine Scott Wood Papers, Huntington Library, Pasadena, especially WD80, WD276, WD277, and WD292. I am indebted to the staff of the Huntington Library for research help and financial support.

CHAPTER 6: Much of this chapter is based on interviews with Alice Anderson's three sisters, Kathleen Ball, Frances Durham, and Claire Fitzpatrick. Kathleen Ball and Frances Durham were interviewed by Mimi Colligan in 1981, and I interviewed Claire Fitzpatrick numerous times between 1995 and 1997. Thanks to Libby Shade, who shared her interviews with me, and to Peto Beal for help in reproducing family photographs. My thanks also to former garage workers Luck Garlick (Johnstone), Marjorie Horne, Pat Peterson, and Marie Edie (Martin's) son Ted Martin. See also Mimi Colligan, "Alice Anderson, Garage Proprietor," in *Double Time: Women in Victoria—150 Years*, ed. Marilyn Lake and Farley Kelly (Melbourne: Penguin Books, 1985), 305–11. My thanks to Mimi Colligan for generously allowing me access to her files.

CHAPTER 7: My thanks to Mrs. Kathleen Howell (nee Gardiner) and to the family of Mrs. Marion Bell, most especially Cedric Bell, for granting me interviews and loaning material that greatly helped me with the research for this chapter. On tracing the circumference of the continent, see Peter Bishop, "Driving Around: The Unsettling of Australia," *Studies in Travel Writing* 2 (Spring 1998): 150. On white women's relationship to Aboriginal women, see Fiona Paisley, *Loving Protection? Australian Feminism and Aboriginal Women's Rights, 1919–1939* (Carlton South:

Melbourne University Press, 2000), 144ff.; Deborah Bird Rose, "Nature and Gender in Outback Australia," *History and Anthropology* 5, nos. 3–4 (1992): 403–25.

CHAPTER 8: Reports of Belcher and Budgell's trip are given in M. L. Belcher, *Cape to Cowley via Cairo in a Light Car* (London: Methuen and Co., 1932); Margaret Belcher and Ellen Budgell, "That Other Great North Road: Eight Thousand Miles from South to North of the Dark Continent in a Six-Year-Old Oxford," *Morris Owner*, November 1930, 1108–10; "'Bohunkus' Comes Home: Dominions Want British: Cape-Cairo Motorists at Cowley Works: Africa Thrills," *Oxford Mail*, 2 October 1930, 1; "Cape Town to Cairo by Car: Experiences of Two Women Motorists," *London Times*, 17 September 1930, 11; "Perilous Journey Ended: Women Who Drove from Cape to Cairo," *North Berkshire Herald*, 3 October 1930; C. R. Lucato, "From Cape Town to Oxford in a 1924 Morris-Oxford," *Morris Owner*, November 1930, 1105–6. My thanks to Trevor Williams of the Cowley Local Historical Society for research help with the Cowley material. For South African reports of Belcher and Budgell's trip, see *Cape Times*, 2 April 1930, 10; 3 April 1930; 7 April 1930, 9; 12 April 1930, 4; 14 April 1930; 29 April 1930, 9; 5 May 1930, 11; 7 May 1930, 5; *Diamond Fields Advertiser* (Kimberley), 8 April 1930. My thanks to Jackie Loos of Cape Town for research help on South African material. Margaret Belcher, *Cape Town to Oxford via Cairo, 1930*, MS held in the Rhodes House collection, Bodleian Library, Shelfmark MSS Afr. s. 277; "Reminiscences of a Social Welfare Officer, Eastern Region, Nigeria, 1948–1958," Rhodes House collection, Bodleian Library, Shelfmark MSS Afr. r. 215; Cape Town Society for the Protection of Child Life, *Annual Reports*, nos. 21–26, 1928–33; "Letters Home of a Social Welfare Officer, Calabar, Nigeria, 1948–1958," Rhodes House collection, Bodleian Library, Shelfmark MSS Afr. s. 1343. My thanks to Sue Mathews for help in researching this collection.

On Cape-to-Cairo and the British imagination, see Lois A. C. Raphael, *The Cape-to-Cairo Dream: A Study in British Imperialism* (New York: Columbia University Press, 1936), 21–23; Peter Merrington, "A Staggered Orientalism: The Cape to Cairo Imaginary," *Poetics Today* 22, no. 2 (2001): 323–64.

On the Cape-to-Cairo automobile trips, see James B. Wolf, "Imperial Integration of Wheels: The Car, the British, and the Cape-to-Cairo Route," in *Literature and Imperialism*, ed. Robert Giddings (New York: St. Martin's Press, 1991), 112–27.

Index

Aboriginal Australians: suppression of, 105, 135–36, 137, 138, 139; white women's response to, 123, 128, 131–32, 137, 139
Aboriginal Protection League, 136
Adas, Michael, 144
Adelaide Advertiser, 123
advertisements, 36–38, 41, 45, 75, 76, 82, 83, 110, 137, 138
aero-engine construction, 53, 57
Africa, journeys in: Cape de Cairo, 141, 145–46, 148–57, Cape to Cowley, 140–142, 148
African Americans: and car ownership, 6; as chauffeurs, 27–28; and Jim Crow provisions, 6, 87; and suffrage activism, 87, 89, 102
Africans: romanticized as "noble primitive," 152–53; white women's positioning toward, 141, 153, 154–56
Albert Kahn Associates, 49
Alice Anderson Motor Service, 107, 110, 111–14, 115–16, 117, 118–19, 131, 162
Allen, Grant, 117
Allender, Nina, 102
Allies Field Ambulance Corps, 34
Alpha Club, Chicago, 89
Amalgamated Engineering Union (AEU; UK), 55, 60
Amalgamated Society of Engineers. *See* Amalgamated Engineering Union
American Automobile Association, 22
Anderson, Alice: background of, 108; death of, 107–8; education of, 108; and first car, 108–9; garage business of, 107, 109, 110–19; inventions of, 112; sexuality of, 117–18; and vision of professional motorists, 111, 166

Anthony, Susan B., 92
Aria, Eliza Davis, 20, 24–25
Arrol-Johnston motorcar and company, 47, 55; decline of, 57, 59, 60–61; Heathhall factory, 49, 60; Tongland factory, 48–51, 52, 53, 54–55, 56–59, 60, 61. *See also* Galloway Engineering Company; Galloway motorcar
Artemis (pseudonym), 20, 22
Ashberry, Annette, 62
Atalanta Ltd., 62
Australia: automobiles and unification of, 104, 105–6; circumnavigation by motorcars, 120, 122–24; frontier insecurity of, 104, 105, 135; motorists and nation building in, 107, 120–21, 134–35, 137, 139, 140, 159; population of, 105; race relations in, 135–39; and women's perceived difference from "old world" women, 106–7; women's position in colonial, 123, 134, 137–39
Australian Automobile Trade Journal, 113
Australian Motorist, 20, 110, 112
Australian Women's Army Service trainees, 164
Autocar, 12, 31–32, 36
auto engineers. *See* Galloway Engineering Company
auto industry, UK: and class, 46, 49, 51, 57, 60, 61, 62, 63; during WWI and postwar period, 46, 54, 61, 150–51
auto industry, U.S., 66–67
auto mechanics: in Australia, x, 107–19; in UK, 32, 33, 39–41, 42–44; in U.S., 69, 70–72; as working-class male trade, 35, 70, 118

automobile: culture of, 11, 12; gasoline, 13, 15, 83; history of, 13, 166–67; as term, 167. *See also* cars
Automobile Chamber of Commerce (Aust.), 112
Automobile Club of Victoria, 131
automobile sportswomen, 22, 23, 113
automobile technology: and Australian national unification, 104, 105–6; changes during interwar years in, 161; designed for women's perceived weaknesses, 83–84, 161; drivers' and passengers' learning of, 13, 16, 32–33; as legitimating white superiority, 136–39, 143, 144, 146, 154, 155, 156, 157; men's exclusive claims over, 12, 119; as statement of national allegiance, 150–51; as symbol of American superiority, 141, 143; as symbol of global inequality, 141; as weapon against working-class militancy, 146; and women's awareness of masculine privilege, 12
automobiles. *See* motorcars; *specific cars*
automobilists, private, 14
automobility, democratization of, 28, 35, 64, 66, 67, 83

Baden-Powell, Robert, 152
Baldwin, Captain Thomas S., 22
Batten, Jean, 122
Beattie, Eleanor Lorraine, 68
Behling, Laura, 65
Belcher, Margaret, 140, 141, 142, 147–48, 149, 150, 151, 152, 153, 154–57, 166
Bell, Miss Marion, 123, 125, 130, 138
Bell, Mrs. Marion, 122–23, 125–26, 127–28, 129, 130, 131, 135, 137, 138
Bell, Norman, 126
Benedict, Crystal Eastman, 78
Benz family, 1
Birmingham Jubilee Singers, 148
bodily experience of gender differences, 4, 9–10, 17–18, 20, 21, 26, 27, 46, 55, 69, 158
Bohunkus (motorcar), 142, 148, 149, 151, 152, 153, 154, 156, 157
"Bohunkus and Josephus" (song), 148–49
Boissevain, Inez Milholland, 78

books, children's, 106
Borthwick, Gabrielle, 32, 33, 39–40, 166
Borthwick, Lord, 42
Borthwick Garages Ltd., 43
Boston YMCA driving school, 13
Bournville model village (UK), 55
Bridge, Miss, 61–62
British colonialism, 144–57; and British adventurers in Africa, 1920s, 145–46, 147, 153–54; downfall of, 147; and Girl Guides, 152–53; in sub-Saharan Africa, 144–45; and white South African motorists, 1930s, 140–41, 147–50, 151–52, 153, 154–57
British raj, 147
Brooklands racing circuit (UK), 25
Bruce, Mary Grant, 106
Budgell, Ellen, 141, 142, 148, 149, 152
Bulkley, Nora, 32
Bull, E. F., 62
Burberry, 35, 38
Burke, Alice Snitjer, 82
Burlingame, Margaret R., 83, 84
businesses, motoring: in Australia, 107–19; in UK, 32, 33, 39–41, 42–44, 70; in U.S., 68–72

Campbell, Edwin, 48
Campbell, Margaret, 79
Campbell von Laurentz, Baroness, 20
Cape-to-Cairo: The Record of a Historic Motor Journey (Stella, Court Treatt), 145
Cape to Cairo journeys, 141, 145–46, 148, 149, 150, 151, 152, 153–57
Cape to Cowley journey, 140–42, 148
Cape to Cowley via Cairo in a Light Car (Belcher), 148, 157
Cape Town Society for the Protection of Child Life, 151–52
Careers and Openings for Women (Strachey), 45
Car Illustrated, 12, 19, 20
car manufacturing. *See* auto industry, UK; Galloway Engineering Company
car ownership: in Australia, 106; in South Africa, 150–51; in UK, 2, 35; in U.S., 2, 35, 64–65
cars. *See* motorcars

INDEX

Carstairs, Barbara "Joe," 42, 43
Caswell, Mrs. L. W., 71
Charlesworth, Mrs., 32
Chartered Institute of Professional Engineers (UK), 62
chauffeurs, 13, 16, 27, 28, 31, 32, 38, 39, 42, 43, 68–69
Christie, Stella, 126, 127, 130, 131, 132, 133
circum-motoring Australia, 120, 122–24
Citroën motorcar, 125–26
class, in Africa: and British upper-class travel, 145–47, 149, 151, 153–54, 157; and colonial middle-class women, 140–41, 145, 151, 152, 153, 157
class, in Australia: and middle-class motoring careers, 108, 111, 113; structure of, 111; and superiority, 118–19
class, in UK: and auto engineers, 46, 49, 51, 57–58, 60, 61, 62, 63; and conflict, 146, 147; and employment, 28–29; and mechanics as working-class male trade, 35, 70, 118; and professional motoring, 30, 32–33, 35, 36, 41; and upper-class motoring, 20, 21, 24, 64
class, in U.S.: and motorists, 13–14, 64, 65, 66, 67; and professional motorists, 16, 27–28, 72; of suffragists, 70–71, 73, 80, 102
Cleveland Leader, 82
Clincher Motor Tyre Company, 34
College Equal Suffrage League of Northern California, 80
Collier's, 91
Collingham, Elizabeth, 147
Collins, Betsy, 31
colonial femininity, 111, 134, 151, 153, 154, 155–56
colonialism, 141, 143–44. See also British colonialism
Common Cause, 35, 39
Congressional Union for Woman Suffrage (US), 79, 82, 89, 90, 91, 92, 93–94, 96, 97, 98, 99, 101
consumers, 25; culture of, 64, 67; influence on car design by, 83–84, 161
consumption: and colonialism, 141, 143; and social change, 5–6, 28, 35, 64, 66, 67, 72; women's aspirations framed by, 65–66, 72, 78, 82–83, 85, 87, 100–101, 102, 159
Court Treatt, Major, 145–46, 149, 150, 153–54
Court Treatt, Stella, 145–46, 148, 149, 150, 151, 153, 157
Cowan, Ruth, 162
Crook family, 132–34, 135
Crozier, Gladys Beattie, 20
Cummings, Ivy, 40
Cuneo, Joan, 22, 23, 25

Daily Graphic, 20
Dakar to Maswa journey, 147
Day, J.O.M., 148
Deanne, Nancy, 69
De Dion motorcar, 20
de Havilland, Gladys, 34, 45
Delco garage, St. Louis, 70–71
democratization of automobility, 28, 35, 64, 66, 67, 83
Directory of Victoria, 110
Doan, Laura, 42, 44
Dodds, Johnny, 148
Donovan, Frank, 66
drivers. *See* motorists
driving schools, 13, 32, 112–13, 119

Edge, Selwyn, 30
education: of drivers and passengers, 13, 16, 32–33, 112–13, 119; professional motoring, 33, 39–40, 70, 71, 113–14
Egypt, 156
Ellington, Miss, 40
Ellis, Christobel, 25, 26, 36
employment: men's pressure on women's, 58–59, 62; and professional motorists, 30–35, 38–41, 42–45, 46–47, 68–69. *See also* businesses, motoring; Galloway Engineering Company
Emporia Gazette, 95
engines, internal combustion, 13
Evening News (London), 43
Everett, Jeanette, 68

Faderman, Lillian, 65
fashions, 24–26, 35–36, 38, 44, 77, 114, 115, 130

female masculinity: in Australia, 115–17, 119, 159; as disreputable, 44, 114; in UK, 36, 43–44, 45, 46, 64, 65, 72, 159; in U.S., 65, 72, 78, 85, 159

femininity: colonial, 111, 134, 151, 153, 154, 155–56, 157; during interwar years, 160–61; meanings of, altered by motoring, 5, 8–9, 36, 52, 69, 72, 85, 87, 106, 107, 130, 134, 139, 151, 157, 159; modern white Australian, 106–7, 111, 114, 117, 130–31, 134, 139; U.S. context of, 28, 65, 69, 72, 77, 78

feminism, 7–8; and auto workers, 53–54, 63; and commodity consumption, 65–66, 73, 78, 82–83, 87, 100–102, 159–59; during interwar years, 44–45, 160–61; and motorists, 46; second wave, 7, 166; and women's employment, 39, 44–45. *See also* Galloway Engineering Company

Field, Sara Bard, 86, 89, 90, 92, 94–95, 96, 97–98, 99–100

films, 76, 106

First Aid Nursing Yeomanry, 34

Fisher, Harriet, 143

Fitzpatrick, Claire, 113

Fleury, Gabrielle, 115

Ford, Henry, 67

French, Anne Rainsford, 13

French-Sheldon, May, 141

Gaines, Mrs. Dan, 67, 68

Galloway Engineering Company, Tongland, UK, 47–63; and class consciousness, 49, 51, 57, 61, 63; closure of, 60; and feminist campaigns, 53–54, 63; founding principles of, 48, 63; as industrial relations experiment, 55–56; men's pressure on women's employment at, 58–59, 62; women workers at, 49–54, 56–59; working conditions at, 52–53

Galloway motorcar, 46, 47–48, 60

Galloway Motors Ltd., 46. *See also* Galloway Engineering Company

garage women: in Australia, x, 107–19; in UK, 32, 33, 39–41, 42–44; in U.S., 69, 70–72. *See also* Anderson, Alice

Garlick, Lucy, 113

gender. *See* sex/gender distinction

Gentlewoman, 20, 51

Girl and the Motor, The (Ward), 15, 17

Girl—A Record and an Oldsmobile, A (brochure), 74, 75

Girl Guide movement, 152–53

Give a Girl a Spanner (slogan), ix, 166

Glasgow Institute of Technology, 51

Glasgow School of Art, 51

Glasgow Technical College, 62

Glidden, Charles, 143

Glidden, Mrs., 143

Graphic, 35, 48

Great Depression, 147, 148, 160

Griff, C., 32, 39, 43

Hall, Mary, 141

Hall, Radclyffe, 41, 44

"Her Wheel" (column), 112

Hitchcock, Mrs. A. Sherman, 18, 19, 68, 78

Hodgson, Frances, 40

Hollywood, 73–74, 76

Holmes, Helen, 73

Holmes, Vera, 51–2

Horne, Marjorie, 113, 115

Horseless Age, 12

Howell, Kathleen, 131–32, 134

Hudson Essex Six motorcar, 126

Hulme, Kathryn, 4

Hupmobile motorcar, 108

Ibbotson, Miss, 43

industrial relations, in UK, 54–56, 58–60, 61, 146, 147

Institution of Automobile Engineers (UK), 56

Jeune, Lady, 19, 24

Joan of Arc, 117

Johnson, Amy, 122

Joliffe, Frances, 90, 92, 94, 95

Jones, Rosalie, 71–72, 79, 80, 81

Kansas City Post, 98

Kellerman, Annette, 73

Kelsey, Major, 145

Kennard, Mrs. Edward, 19, 21

Kent, Mrs., 99

Kindberg, Maria, 86, 92, 93–94, 95, 96, 97, 99, 101, 102
King, Anita, 73, 74, 76, 77
King, Jessie M., 51
Kinstedt, Ingeborg, 86, 92, 93–94, 95, 96, 97, 98, 99, 100, 102, 166
Kirkcudbright, UK, 48, 51, 52, 58
Kirkcudbright Advertiser, 58
Kissel Motor Car Company, Wisconsin, 74, 76
Klein, Rose, 71

Ladies' Automobile School, London, 32, 39
Ladies' Automobile Workshops, 32, 33, 39, 40
Ladies' Realm, 20
Lady's Pictorial, 40, 49
Lancia Lambda motorcar, 131
Lees, Miss, 61
lesbianism, 41–42, 44, 65, 115, 117–18, 119
Levitt, Dorothy, 20, 21, 24, 25, 30, 166
Licensed Vehicle Workers' Union (UK), 39
Limit (magazine), 52–53, 56, 57, 61
Lincoln Highway Association, 91
Literary Digest, 88
Lockwood, Katherine, 68
Locomobile motorcar, 13, 143
London Cab and Omnibus Company, 31
London Sphere (magazine), 18
Loughborough Technical College, 62

MacKaye, Hazel, 92
Maranboy Aboriginal Reserve, 137
Margaret Partridge and Company, Electrical Engineers, 61
Marion Bell's Char-a-Banc Service, 126
Martin, Marie, 113–14
massacres, 136
Maxim, Hiram Percy, 13
Maxwell Car Company, 74, 78
Mayo, Miss, 40
McOstrich, P., 41
media representations of motorists, 17–18, 36, 67–68, 77, 122, 123, 125, 127, 129–30, 135, 147, 158, 162, 166. *See also* films; *specific periodicals*

men: attitude toward professional women motorists, 31–32; attitude toward women drivers, 2, 11, 14–15, 17–18, 21, 66, 161–62, 164; and characteristics attributed to women, 21–22; and early racing fashions, 25
Minerva (pseudonym), 20
Ministry of Munitions (UK), 54, 55
modernity, bodily experience of, 9–10, 12–15, 22, 23–24, 44, 63, 64, 158, 159
Moehle, Jean Earl, 78
Monte Carlo Rally, 131
Morris Motor Company, 151
Morris Oxford motorcar, 141, 148
Motor, 18, 64
Motor Age, 141
motorcars: Citroën, 125–126; electric, 13, 14–15; De Dion, 20; Galloway, 46, 47–48, 60; Hudson Essex Six, 126; Hupmobile, 108; Lancia Lambda, 131; Locomobile, 13, 143; Morris Oxford, 141, 148; Napier, 143; Oldsmobile Six, 123; Overland, 74, 75, 100, 101; Saxon, 82
Motor Drivers' Employment Agency, 31
Motor Field, 78
motoring: bodily adaptations to, 4, 10, 20, 21, 26, 27, 100, 158, 159; and gendered differences, 3–5, 7–8, 10, 15–18, 26–27, 66, 84, 158, 159, 167; historical perspective on, 166–67; for households, 162–63; as political action, 71–73, 78–83
"Motoring for Ladies" (*Ladies' Realm*), 20
motorists: attitudes toward male drivers, 22, 27; challenges to restrictions on, 9, 11, 23–24, 158; colonial travel narratives of, 128–29, 137, 139, 142, 145, 147–48, 152, 153, 154, 155, 156, 157; and electric cars, 14–15; male (*see* men); political quest of, 8–9, 90; public scrutiny of, 24; "woman motorist," 2; writings of early, 4, 9–10, 12, 15, 17, 18–22, 24–25. *See also* professional motorists
Motor Maniac, The (Kennard), 19
Motor Trade Association (UK), 40
motor vehicle manufacturing. *See* auto industry, UK; Galloway Engineering Company
Mudge, Eva, 67

Munitions Act (UK), 54
My Motor Milestones: How to Tour in a Car (Campbell von Laurentz), 20

Napier motorcar, 30, 143
National American Woman Suffrage Association (NAWSA), 79, 82, 83, 89, 96
National Council for Women (UK), 57
National Union of Women's Suffrage Societies (NUWSS), 38, 39, 53
Nebraska State Journal, 99
Neil, Margaret, 71
neurasthenia, 22
Neville, Alice Hilda, 32, 43
New Woman communities, 110, 114
New York–to–Paris race, 143
New York Tribune, 86, 98
Northern Territory Times, 127

Official Guide to the Lincoln Highway, 91
Oldsmobile Six motorcar, 123
Olds Motor Works, 74
O'Neil, Sheila, 30–1, 39
Outing (magazine), 17, 19, 22
Overland motorcar, 74, 75, 100, 101

Panama Canal, 90
Panama-Pacific International Exposition, 74, 76, 90–92
Pankhurst, Emmeline, 32, 52
Parbury, Miss, 40
Paris-to-Peking race, 143
Paul, Alice, 79, 82, 92, 95, 96–97, 98
Phillips, Amy, 75
physicality, 9–10, 14, 16, 19, 25, 26, 45, 128–29, 130, 155
Post, Emily, 91
press. *See* media
Preston, Amelia, 32
Preuss, Amanda, 74, 75, 77
professional motorists: aspirations of, 111, 162; in Australia, 105, 107–19; in U.S., early 1900s, 67–68, 69, 70–72; in UK, during WWI, 30–35, 40–41, 42–45. *See also* Anderson, Alice; garage women
Providence Sunday Journal, 99

Pullinger, Dorothée, 49, 62–63, 166
Pullinger, T. C., 48, 49, 54, 55, 56, 60
Pursell, Carroll, 67

Queen, the Ladies' Newspaper and Court Chronicle (UK), 32, 47, 48

Race, The (film), 76
race-based superiority: in Australia, 105, 128, 131–32, 135–36, 137, 139; in colonial Africa, 152–53, 154, 155–56, 157; in U.S., 6, 27–28, 89, 102–3
races, automobile, 143
railways, 104, 105, 144–45
Ramsey, Alice Huyler, 74, 76–77, 78, 166
reliability trials, 74, 78, 131
Remy Car Service, 40
Restoration of Pre-War Practices Act (UK), 58
Rhode Island Equal Suffrage Association, 99
Rhode Island Society for the Suppression of Vice, 99
Richardson, Nell, 82
Ritza, Madelaine G., 84
Robertson, Jean, 113, 131
Rollins, Montgomery, 21
Rowbotham, Dorothy, 53, 57, 60, 61
Royal Automobile Club (UK), 40
Royal Naval, Military and Air Forces tournament, 34
Russey, Wilma K., 69

Sackville-West, Vita, 36
Salmon, Balliol, 36
Sandford, Gladys, 126, 127, 128–29, 130, 131, 132, 133, 135, 137, 163, 166
Sandford, John, 127
Saxon motorcar and company, 82
Scharff, Virginia, 66
Schultz, Olive, 69
Scott, Blanche Stuart, 74, 75–76, 77–78
Scottish Women's Hospital, 34, 39
Searchlight (journal), 99
Self, Mrs. E. M., 70–71
serial queen genre, 73–74
sex/gender distinction: and automobile technology, 4–5; and bodily experience,

4, 9–10, 17–18, 20, 21, 26, 27, 46, 55, 69, 158; and negative stereotypes, 7, 8
Shaw, Anna Howard, 83
Shaw, Nate, 6
Singleton Station, Australia, 132, 135
social structures. *See* sex/gender distinction
Society of Women Motor Drivers (UK), 39, 45
Society of Women Welders (UK), 39
So Simple That Even a Woman Can Drive It (slogan), 2
South Australian Register, 138
Sphere (magazine), 37
Star Engineering Company, 147
steam-powered automobiles, 13
stereotypes. *See* sex/gender distinction
Stewart-Warner (manufacturer), 38
Stockwell, Ruth H., 71
Strachey, Ray, 39, 45
Strickland, Diana, 147, 148, 151, 154, 157
Strike, General, 146
suffrage movement: in Australia, 106, 107; and automobile campaigns, 5, 86–87, 88–90, 92–103; and black women, 89, 102, 103; and march from New York to Albany, 79; in UK, 38, 53, 54, 65; in U.S., 71–73, 78–83
Suffragist (newspaper), 96, 102
Sunset Magazine, 71
Susan B. Anthony Amendment, 92
Sydney Mail, 125

Taking the Wheel: Women and the Coming of the Motor Age (Scharff), 66
Tatler, 20
taxi drivers, x, 30–31, 43, 68–69, 152
Teape, Minerva, 4
Teape, Vera, 4
Technology and Culture, 9
Terborg-Penn, Rosalyn, 89
Tongland, UK, 47, 48. *See also* Galloway Engineering Company
Town and Country, 13
transcontinental journeys: and Aboriginal people, 123, 128, 131–32, 135–36; in Africa, 140–41, 142, 144–57; media reports of, 122, 123, 125, 127, 129–30, 135; and motorists as heroes, 122; and narrative focus of motorists, 128, 155–56; as nation-building exercises, 120–22, 140; political purpose of, 86–87, 90, 92–103; sponsorship dearth for, 121; symbolic importance of, 133–34; in U.S., 74–79, 82
transnational journeys: colonial context of, 140, 159; popularity of, 143; routes of, 143–44
Troubridge, Una, 44
Turner, Dora, 62
"Two in a Car: Women Wanderers in the Waste" (review), 122

unions, UK: and campaigns against female employment, 56, 58–59, 60, 61, 62; post-WWI, 39, 48, 54, 55, 56, 146

van Buren sisters, 69
Vernon, Mabel, 92, 95, 96, 97
vitalist theories, 22
Vote (journal), 39

Walker, May, 36
Ward, Hilda, 15–17, 18, 19, 21, 26–28, 72, 166
Warland Dual Rim Company, 38
Warner, Joshua, 125, 126, 137
Warwick School of Motoring, 32
Washington Herald, 90
Well of Loneliness, The (Hall), 41, 44
Wheels for a Nation (Donovan), 66
Whitcombe, Mrs., 43
White, Pearl, 73
White Car Company, Cleveland, 83
White Service Steam Laundry of Croydon, 63
"Why Women Are, or Are Not, Good Chauffeuses" (*Outing*), 17, 18
Willys-Knight dealership, Iowa, 71
Willys-Overland Company, 74, 75
Wilson, Woodrow, 89, 92, 94, 95
Wolverhampton Star motorcar, 147
Woman and Her Car, The: A Chatty Little Handbook for All Women Who Motor or Who Want to Motor (Levitt), 20–21, 30

Woman and the Motor Car: Being the Autobiography of an Automobilist (Aria), 20, 24
Woman Engineer (journal), 39, 40, 43, 62
Woman's Motor Manual, The: How to Obtain Employment in Government or Private Service as a Woman Driver (de Havilland), 34, 45
"Woman's Point of View, A" (column), 20, 36
"Woman's Viewpoint of Motoring, A" (column), 18–19
Woman's World (magazine), 108, 112
Woman Voter (magazine), 99
Woman Who Did, The (Allen), 117
"Woman Who Does, The" (article), 117
"Women A-Wheel" (column), 20, 22
Women's Automobile Clubs (US), 78
Women's Emergency Corps (UK), 33
Women's Engineering Society (WES; UK), 38, 52, 55, 59–60, 62
Women's Industrial League (UK), 38
Women's Industrial News, 39
Women's Legion (UK), 33, 36
Women's Motoring Club of New York, 74
Women's Political Equality League, 99
Women's Service Bureau (UK), 38
Women's Social and Political Union (UK), 54
Women's Volunteer Reserve (UK), 32, 33, 34
Wood, Charles E. S., 95
World War I: impact on women, x, 35–36, 39, 41, 47, 63, 69–70, 86–87, 115, 159; impact on women motorists, 32–35, 37–40, 46–47, 110; and postwar working conditions, 44–45, 58–59; social impact of, 64
World War II, ix, 163–66

X Garage, 42–43